Simply Three Words

to Mindfulness

Simply three words to Mindfulness

Transform your thoughts, relationships and lifelong journey through mindfulness and meditation

Alex Bruce

Simply Three Words to Mindfulness

Transform your thoughts, relationships and lifelong journey through mindfulness and meditation

Alex Bruce

ISBN-13:978-1-7908-8091-1

Dedicated with love to
Uncle Al

May you one day know the reach of your ripple

Table of Contents

Foreword

The more mindfulness that one finds, the more mindful one wishes to become.

Every day our nervous systems are under attack. At best it's a low-grade burn and at its extreme it's a full-fledged assault every waking minute and even in our sleep. Society's expectations to do more with less and respond-this-instant to emails, texts and calls have made it difficult if not impossible to slow the battering. Not surprisingly, in a lot of cases our worst and most dangerous enemy is ourselves.

We are controlling our inner world through our thoughts and study after study reveals the same information; that the majority of our thoughts are heavily weighted in the negative arena. This reality and the undue stress that it causes is proving to be incredibly harmful for us.

These statements may seem provocative however just take a look at our friends, coworkers, neighbors and even in our own homes and you'll see the stress, anxiety, depression and even burn-out that is plaguing so many of us. Ironically, feeling utterly alone in despair and struggle is something that actually shows how connected we truly are. There is great hope and it's easier than we'd think.

Any one of us can practice mindfulness in one moment because a moment is all it takes. What's really exciting is that every single moment of mindfulness that we experience exponentially strengthens our nervous systems. The more we practice, the healthier we get. It's that easy.

If you could change your entire outlook on life with simply three words, would you do it?

Imagine a complete paradigm shift, a totally new inner world, a powerful way of thinking differently and all of it through the use of a phrase or a sentence. Simply three words - that can be all you need to completely change your life. The words may entice you to look at something with entirely new eyes. They may motivate and inspire you or perhaps they will ground and center you. Whatever they are and whatever they do, they can help to give you the life that you want and the future that you didn't know you were looking for.

For so many of us, we go from stimulus to reaction. This equates to someone cutting us off in traffic and us leaning on the horn - it's a no-brainer...one equals the other. It may be a certain aggravating behavior from a loved one that in turn begets a snide remark from us. Of course, that's how it goes - they know that it drives us crazy when they do that so yes, their behavior leads to our insult. Stimulus and reaction - that's how so many of us live, if not all of the time then at least some of the time.

Mindfulness is actually the spot in the middle of stimulus and reaction that most of us weren't even aware of. It's imperceptible in the beginning and then gets larger the more we use it...much like our biceps when we start lifting weights at the gym.

The phrases contained in this book are intended to help you increase your awareness in between stimulus and reaction so that stimulus-reaction magically becomes stimulus-<u>mindfulness</u>-*response*. That's right, the mindfulness right after the stimulus miraculously changes our reaction (habituated return) to a calmer, more thoughtful (more centered) response.

How different would your life be if you responded versus reacted even five percent of the time more than you do now? What harsh words would not have been spoken? Which embarrassing or humiliating actions would not have occurred? How many less apologies would we have had to make?

Let's say that your preferred phrase out of this entire book was, "Stop and breathe", and you actually used it when you were triggered - who would you be? Imagine that every single time you felt the blood rush to your face, your palms start sweating or your heart feel like it's about to explode you were mindful enough to remember, "Stop and breathe", and then you did. What would that look like?

It's not impossible to imagine three-word phrases sticking with us, echoing in our minds. Consider just a few that have already stuck in our heads as a reminder of what they represent. From Bugs Bunny asking, "What's up, Doc?" and Porky Pig telling us, "That's all, folks!" as children to Nike inspiring us to "Just do it."

You can tell a good life story using three words at a time: I love you; please marry me; yes, I do; Honey, we're pregnant; the terrible two's; Honey, pregnant again; a family vacation; some family drama; lots of love; finally, empty nesters; gift of grandchildren; some life wisdom; Goodbye old friend.

So, if we can tell an entire good life story using three words, why not make your good life story even better by using some of the three words contained in this book.

What if you lived your life three words at a time: wake up happy; love my family; eat something healthy; smile and wave; go to work; make good decisions; drive home safely; make someone's day; relax and unwind; do your part; spend time privately; communicate and laugh; best sleep ever.

Looking back over the years, some of the most pivotal moments, thoughts and beliefs in our lives employ simply three words. Today it's time to introduce three more.

Everyone these days is talking about Mindfulness. People either have it, they want it or they want to know what it is. To many it's an elusive, confusing and mysterious concept. Simply defined, mindfulness is being completely aware in this very moment. The emphasis is on the words "completely" and "aware." That's it. It's about slowing down and being present right here, right now because this moment is where life actually occurs. Being mindful is you being immersed in reading this book and feeling your body sitting or standing as you read it. It's about the awareness of the sensation of the book in your hands. It's feeling the breath in your lungs and about allowing your muscles to relax. It's the awareness of each and every emotional reaction that you have while you're reading it.

How is that so different than what we do every other minute of every single day? Well, for most of us we're also thinking about what we're going to do for dinner tonight, how angry we still are from that argument this morning with our family member or some other experience of "then", either in the future or the past. If we're also concentrating on "then" it means that we're missing the full richness of experiencing "now."

In my practice with clients I like to describe and sum up Mindfulness in simply three words; compassion, curiosity, consciousness. Later chapters will go in to more detail but basically it entails:

1) Compassion for everyone, especially for ourselves. How different would this day look if you were being compassionate with everyone including you - understanding, kind and supportive versus possibly negative, critical or harsh?

2) Curiosity of what is happening in this moment - how we're feeling and thinking without the narrative or judgement that usually instantaneously proceeds.

3) Consciously slowing down and being aware of what's truly transpiring. This is the exact opposite of being on auto-pilot as we've been trained to do our entire lives in order to become better at multi-tasking.

We like to think that we're always conscious yet how many times have we left the house only to question whether we turned off the stove, unplugged the iron or locked the door? How many times have you gotten to where you were driving, walking or transitting and then realized that you can't remember the trip? We've all been there and all of that is OK; realizing that we're not being conscious all the time is the first step to being conscious. Congratulations - you're already on your way to becoming more mindful! And if at this point you can reflect that sometimes you're not very compassionate with yourself or that you react versus respond because you don't feel you have the time to be curious or investigate the full situation then you've just scored a hat-trick and you're well on your way to mindfulness.

The intention of this book is to cause you a moment of pause and reflection regarding your thoughts and beliefs and to inspire you to choose to live your life in a way that is healthier for you and for those around you. These phrases are introduced as an offering for you, in the hopes that they help you to make more peace with our everyday occurrences. I envision it in a place that is easy to access, such as on a coffee table or sitting on the back of the toilet where you can give yourself a moment to read a page or two and digest what you're meant to digest.

My sincere wish for you is that you find countless sayings in this book that stir you to view life in a more positive and mindful manner. I hope that every single one of these phrases or sentences inspires you to look at yourself, others and life in general with a more positive and thoughtful approach. That being said, even if only one of them really speaks to you, isn't it still worth your time?

The background of this book is a bit of a mixed bag, as is mostly everything in this life. Although I've been teaching Mindfulness and Meditation since 2012, the majority of my working life includes well over two decades in Finances, involving coaching and training staff and Managers. Throughout that time, I met with several hundred people on a regular basis over an extended period of time. Many of them, in the safety and security of their private offices, would open up to me with their fears, desires and secrets, looking for advice. They talked about their relationships with parents, children, partners and siblings, their jobs, financial concerns and how mental illness was impacting their lives as well as overall life issues. I was tremendously touched and honored every single time that someone trusted me and my recommendations, and I did the best that I could to provide them with counsel that I felt would be in their best interest. I'm grateful for their faith in me and feel that so many of their stories were a reflection of things that a lot of us go through in this life. All of the stories included in this book are true and some aspects, such as names, have been changed in order to respect privacy. My hope is that you are able to connect with some of these reflections, including a number of my own personal sharings, either through your own experience or that of someone you care about and that the words of advice are helpful to you in creating a more mindful, peaceful and joyful life.

<u>Savour this breath</u>

"There is no pleasure in having nothing to do; the fun is in having lots to do and not doing it." - Mary Wilson Little

There is a high likelihood that if you're reading this book right now, you're breathing.

The breath can be compared to an adored family member or a best friend that we tend to take for granted. It's always there for us right when we need it, even on days when it feels like nobody else is. It continuously provides life-giving support and if we allow it to do so, it can encompass us and calm us down in any moment with its presence.

The breath you are taking right this moment has never happened before in the history of this world. It will never happen again. As quickly as it comes in, it fades away and yet just like the endless waves of the ocean, there is another one right behind it. Nature is a beautiful thing and we are all a part of nature.

In this world of go, go, go where faster is better and fastest is best what would it look like for you to actually take a breath and slow down? What would it *feel* like for you to take a breath and slow down?

Feel your breath right now - this gift of life, your best friend. You wouldn't be here without it. What are the sensations that happen in your body to inform you that you're breathing? Can you feel it best in the rise and fall of your belly, the in and out of your chest, the up and down of your shoulders or maybe in the warm and cool sensations in your nostrils?

That breath is gone now and your best friend has given you another. What do you notice with this one? What other sensations are in your body? What sounds do you hear around you? Is there anything for you to smell in this moment? Breathe in this breath as though you are breathing in all of life - because you are. Be curious. And then be grateful.

How does it change things if you close your eyes for your next breath? Do you appreciate it more when you smile with the next one? Can you feel your muscles, your entire body relax with each breath if you slow down and allow it to happen? One after the other, fast or slow, deep or shallow, another breath fills your lungs and nourishes your body. Slow down and allow it to feed your soul as well. Be thankful for this breath, for this moment.

Every so often thinking about the breath being over can arouse feelings of fear or anxiety - one may feel as though time is going too quickly and that's one less breath to breathe in this life. Allow those thoughts to be what they are, and then when you're ready allow them to float away like clouds. Trust in this moment. Trust in this breath and trust that the next breath will come.

In this day and age when we all have to do more with less and the focus is on being smarter, faster and more efficient, take this breath and continue the gift of giving. Intend for this breath to be a gift to your nervous system. Instead of continuing to pump your nervous system full of whatever chemical cocktail we may be living on today (stress, fear, anger, sadness, caffeine to name a few..) take this breath and counteract some of that with nature's gift. Give a little injection of different chemicals in to your body with an awareness of your breath - endorphins and dopamine to start with.

The truth is that life doesn't speed past us when we take a moment to acknowledge our faithful helper, life slows down. Life becomes more delicious. **Savour this breath**

Override the brain

"While conscience is our friend, all is at peace; however once it is offended, farewell to a tranquil mind." - Mary Wortley Montagu

Brains are fascinating. They are incredibly intricate and amazingly delicate works of art. Our brains are here to protect us in ways that we can't even imagine and sometimes they need a little help.

When we are calmly going about our day, most of the time we're strongly engaging the pre-frontal cortex which is the part of the brain in the front of the head that's sometimes affectionately referred to as the "Foreman." It is the most evolved part of the brain and its main responsibility is helping us to keep everything under control. Social sanity often resides here.

Conversely, when our brain perceives any type of threat it immediately engages the amygdala which is one of the least evolved parts of our brain, warmly referred to as part of our reptilian brain. Social insanity often resides here.

Thankfully, the amygdala played a big part of saving our lives as cave-people needing to be able to detect a threat a mile away. Part of its duty is to engage the parts of the body, and required chemical reactions to save us from being attacked and allow us to either attack back or run for our lives. This is where the term "fight or flight response" comes from.

Nowadays, although sometimes we need this response, we are fortunate that we do not need it as much as our ancestors did. Regrettably, often times our amygdala doesn't know that. When we have an ordinary event such as a co-worker looks at us sideways or someone cuts in front of us in line at the movie theatre our amygdala can quickly take over and put us in to fight mode. When this happens our breathing becomes shallow, the heart rate quickens, the eyes dilate, our bodies pump the chemicals cortisol and adrenaline into our bloodstreams and we are fierce and ready to fight. This could lead to a lot of places we don't want to go such as a verbal or physical reaction that we are likely to later regret. The other option is that our amygdala decides that the best course of action is to flee, at which point the breathing is still shallow, our heart rate still quickens and the adrenal glands still pump adrenaline in to our bloodstreams but we get the heck out of dodge and avoid the altercation. As you can deduce, unless we're truly in a threatening position, none of this is especially nourishing for our nervous systems.

The really tricky part here is that when the amygdala kicks in, it overrides the pre-frontal cortex. Yep, like a protective older brother swarming in to help, the amygdala hijacks the most rational part of our brain so that we are much less capable of making a mindful decision. This is good when we're fighting for our lives, not so good if we're arguing with a friend. Have you ever wondered after an episode where you got upset, "What on earth was I thinking?" The answer is simple - you weren't thinking. Lil' ol' Amy took over for you.

Never fear, the answer is near! We can breathe life back in to our pre-frontal cortex (the Foreman) simply by breathing. That's right, when we realize that our brain has been hijacked, through a number of modes such as our crazy thoughts, raging internal and external body and sensations, we can bring our Foreman back within just a few deep breaths. The reason for this is because it overrides your amygdala. You are telling your amygdala that it's wrong. That you are taking long, deep breaths because you are in all actuality, safe. If you were truly fighting for your life, you would not take the time to stop and breathe deeply. This message gets sent to the amygdala and Lil' ol' Amy calms down so that the Foreman can take over and you can now make a more mindful decision. Simple, right? Too bad simple isn't easy.

To strengthen this muscle, try it out when you don't need it as much. If you're slightly irritated over something, take a few deep breaths. And mean it. Don't think about how angry you are while you take them because you're still engaging your amygdala. Be present in your breaths - feel the air coming in to your body and allow your muscles to relax. Take a mini breath vacation. If you watch something on TV or read something in the newspaper or on-line that you find upsetting - you guessed it - take three deep, full breaths. Allow the breaths to fill your awareness as you become fully present in your body. You may find it helpful to close your eyes if it's appropriate - please keep them open if you're driving!

The more that you exercise this muscle to breathe deeply when things cause you to become uneasy, the easier it will become to engage it when you need it most. Remember, a few long, deep breaths to tell Lil' ol' Amy that everything is OK and wait for the Foreman to come back on-line. ***Override the brain***

I am aware

"Most of us take for granted that time flies, meaning that it passes too quickly. But in the mindful state, time doesn't really pass at all. There is only a single instant of time that keeps renewing itself over and over with infinite variety." - Deepak Chopra

The absolute number one step to mindfulness is being aware. In order to feel our breath, we must be aware that we are breathing. In order to feel our anger, we must be aware that we are angry. Awareness is the key to mindfulness. Compassion is what it unlocks.

In any given moment if we remind ourselves, "I am aware", it gives us the opportunity to play a real life "Matrix move" in our day. We can slow ourselves down and take in what is happening. I use this phrase a lot and especially like to use it when I am surrounded by something beautiful. In the past I travelled a lot for work logging about 2500 kilometers a month with a morning drive of anywhere from five minutes to five hours or more. On my longer trips driving in to the Okanogan in British Columbia I would say to myself, "I am aware", and would breathe the moment in. All of a sudden, I would feel myself relax and it was as though my peripheral vision increased - I could soak in more of the mountain and waterfall views. I could feel my hands on the steering wheel and my body resting comfortably in the seat. If my window was open then in a flash, I became more aware of the scent of the trees, the sound of my tires on the highway. It was as though the sky was bluer and more magnificent and the formation of the clouds became so much more interesting. By saying, "I am aware", I allowed myself to take a moment and became more aware. It really does work.

This is something that can be used pretty much anytime, anywhere. Give it a try in a meeting at work and you may be surprised to feel how tense your muscles are or how your thoughts have gone crazy with judgements around what a fraud your co-worker is or some other needless story in your head. Become more aware while you're cooking dinner and listen to the veggies sizzle and the overhead fan buzz, as though they were sounds that you've never noticed before. Become more aware as you look at the face of a loved one and unexpectedly, you're reminded that it was their smile or their eyes that you first fell in love with. When we become more aware of ourselves, others and the world around us we get to enjoy more of what life has to offer. We get to see more, understand more and love more. I am aware

I accept this

"I still miss those I loved who are no longer with me but I find I am grateful for having loved them. The gratitude has finally conquered the loss." - Rita Mae Brown

One of the most painful things we can do to ourselves in this life is to not accept things that are because, we are so emotionally tied to what we want them to be.

Whatever is happening is happening. Whatever has happened has happened. It is what it is. Even if we don't like it or don't agree with it.

If this statement is causing an emotional reaction in you, sit with that for a moment and be compassionate with me and compassionate with yourself. None of us like to accept things that we don't want to accept. None of us. Take a moment and breathe.

When we get caught up in not accepting what is happening it means that we cannot move to our next step. For example, if we refuse to leave the restaurant after our partner has irreconcilably broken up with us, we're not going to find our next partner. If we refuse to leave the building from which our job has just been terminated, we're not going to find our next employer. We can't move on if we don't move on and the first step to moving on is accepting what has happened. If we don't accept it, we are stuck there forever, unable to move.

Many years ago, one of my best friends took his own life. I loved him so much, like a best friend and a brother all wrapped in to one. We understood each other because our backgrounds were so similar and it was as though we both spoke the same language. We just "got" each other.

I was devastated to say the least. And I was stuck in not accepting his death. I was angry and apologetic all at once. I was defiant and in shock. I was not accepting the reality of the fact that he was gone. Every night I would cry myself to sleep and hope that when I woke up it would all be a dream. Every morning I was devastated all over again.

This is an extreme example, but you may find that there's some similarities to this story and to anything in our lives that we refuse to accept.

It took a lot of time and definitely some help, but eventually I accepted that this was true. I accepted that this was happening; I accepted that this was life. There are times that we can changes things and there are times that we cannot. This was one of those times where I could not change what had happened.

When I asked my partner at the time when it would stop hurting, he looked at me compassionately and softly replied, "It may never stop hurting, but it will start to hurt less." I found such comfort and wisdom in his words. He ended up being right.

As time moved on, I accepted that Dave was gone. That acceptance gave me freedom. I was free to have new experiences in my life and to meet new friends. I was free to remember Dave in all of his silliness and free to love his memory versus fighting against his death.

Now I try to use acceptance for everything. When I am losing my mind because I'm waiting in line at the bank and I have things to do, I think to myself, "I accept this", and I calm down. When I see something on the news that makes me cry and I wish that wasn't the way of the world I think to myself, "I accept this." Accepting it is not synonymous with agreeing as there are a lot of things that I don't agree with and I will do everything I can to change what I can. But I mindfully accept the things that I cannot change so that I can mindfully move on and affect what I can. I accept this

Simplify your life

"All of the animals, excepting man, know that the principal business of life is to enjoy it." - Samuel Butler

The power to make life simpler is within us at all times. We may not be able to choose everything that happens in life, but we absolutely have the control to choose how we perceive it and how we choose to respond. We've touched on how we can change our thoughts about what happens, and mindfulness is also about making nourishing decisions physically.

One day a hilarious client of mine named Maggie sat down with me and said, "Sorry I'm late - it's been a hair-straight-back kinda day!" I had never heard that term before and laughed so hard at the mental picture. As time went on, I came to the realization that I had created for myself a hair-straight-back kinda life. It may have made me feel important but it sure didn't make me feel good.

As my mindfulness practice continued, I realized that I was tired of being tired. I would get up between 4 and 5 AM every single morning and after my meditation and yoga I would run like stink until I fell into bed some time after 10:00 just to get a few hours of sleep before I did it all again. My body, my mind and my soul were all in agreement - I was running myself ragged and I was exhausted.

As soon as I realized that the only person who could slow things down for me was me my life began to change. It wasn't my boss that was going to tell me I was working too hard, the charities I supported sure weren't going to tell me to slow down and friends and family were happy to have time with me and my family every single time I was willing to share it. By the time I was finished slicing up my day there was absolutely nothing left for me.

Much as our egos would tend to disagree, there is nothing mindful about living a hectic lifestyle. So many of us wake up to our alarm clock and hit the ground running, going about our day in a panic of multi-tasking just so that we can come home to continue the mania and then collapse into bed knowing full-well that the next day will be much of the same.

It may not feel like it, but living this way is a choice that we have made. I offer you the suggestion to simplify your life anywhere and everywhere that you can. Know that there is a different choice to be made about how you live your life and that the choice is yours to make.

I thought I loved the busy days of over-working, over-volunteering and over-scheduling myself. I received "the talk" from my partner many times about slowing things down but I convinced myself that it wasn't possible.

Thankfully, one day there was a life-altering event at work in which I was passed over for a promotion. I had been running a double workload due to employees departing and Maternity Leaves and I had been promised time and again that my workload would return to normal when we hired more Managers. The promotion became available and it was only during the panel interview that I was informed that this position would be in addition to everything else I was doing because I had handled the extra work so well. I told them that I was already working fourteen-hour days and that there was no possible way I could add more to my plate. I said that my relationships with my clients meant a lot to me and that I wouldn't jeopardize them by diluting my time. At that point, I requested to have my name withdrawn from the process, knowing that I just couldn't handle more.

A week later I received an email with the notice of congratulations to the successful candidate and communication regarding the fact that my account load would increase in order to take on more from the person who received the promotion.

Something inside of me broke that day.

I decided from that point on that eight hours a day was all the organization was going to get from me. All of the extras that I was doing for my clients on my own time stopped. The part that I found was amazing was that although they asked about the reduction in communication, they fully understood when I told them that I just couldn't fit it in to my schedule anymore. I couldn't believe it, but the world didn't end.

It hit me like a tonne of bricks when I realized that I had been doing all of this to myself. I then chose to excuse myself from some of the charities that I was volunteering with because there were other people that did have more time and would be able to dedicate more resources than I was. I allowed my children to quit martial arts because neither one of them could stand it but I had diligently taken them twice a week in the hopes that it would give them structure and skills necessary in this world when they're older. They were thrilled and have since found their own passions, and in the meantime, I opened up huge pockets of time for myself.

The simplification continued - my house didn't have to be as ridiculously clean as I had been keeping it. I didn't have to say yes, every time someone asked me to go out socializing. We started having holiday dinners at home with just the four of us instead of all of the stress that can sometimes accompany large family gatherings. I gave myself permission to simplify my life and it felt incredible. I was the one in charge of releasing my own self-imposing shackles and the freedom was unlike anything I had ever had in my adult life.

Give a little thought as to some of the things that perhaps your sense of obligation or sense of ego are doing to make your life more difficult than it needs to be. Remember, we are human *beings* and not human *doings.*

Allow yourself to do less and to simply be. Simplify your life

Tone it down

"A story is told as much by silence as by speech." - Susan Griffin

As we meditate, we literally strengthen the part of our brain that is responsible for self-reflection. That means that we find it easier to take a look at ourselves as we truly are and as we present ourselves to others.

When I talk about this in groups, I often get comments from people that being more aware of how we come across is a scary thought. That's such a mindful comment to make because it brings attention to the fact that we're already at least partially aware that there may be some room for improvement. That realization is a huge step!

Most of us have a habit or trait that could use some toning down. I worked with one client who has trouble with compulsively cleaning her house and another who consistently put herself down. I have a tendency to speak loudly even if it's just a small group. A peer of mine drives very aggressively without even recognizing that he's tailgating. A co-worker of a friend has a tendency to be hostile, including to his wife, without even meaning to be. All of these are things that could be brought to our attention and perhaps altered with our thought to "Tone it down."

One of my clients is a lovely lady with a huge heart. She also has the propensity to speak non-stop which can be taxing for her peers. It would make me so sad when she would go off on a tangent and her co-workers would roll their eyes, make comments or tune her out because she had passed the point of making her point. I would often look at her, begging her in my head to simply tone it down a bit. What she had to say was beautiful and it all came from her desire to help but what ended up happening was that she would alienate herself from others with her compulsion to speak.

None of us are perfect. That's what makes life so much fun and makes us all so unique and different. Most of us have something (or a couple of somethings) that could serve us better if we turned down the volume just a little and learned to relax, and stop what we're doing. I have found time and time again that when I tone things down, I get a better response from others and also from myself. Catch yourself, remember the phrase and then change your behavior. *Tone it down*

Eliminate the box

"Re-examine all you have been told...Dismiss what insults your soul." - *Walt Whitman*

Most of us have been brought up in boxes whether we're aware of them or not. Mindfully acknowledging the boxes that we're in is the first step to getting out of them. If you are a male then perhaps you've been placed in a box that includes such phrases and beliefs as "Be a man!", "Only sissies cry", "Tough it out" or "Be a winner!" We accepted these offerings in to our boxes without question but what do they even mean? If you identify as a male, then aren't you a "man" already? "Only sissies cry"? No, healthy beings cry. Crying is a way to help us cleanse our bodies both physically and emotionally - male or female. "Tough it out"? There is a reason that there have been huge campaigns in North America urging men to go the doctor and get "checked out" before they "check out" - it's because sometimes "toughing it out" costs people their lives. Interesting that the campaign was specifically targeting males. "Be a winner." this makes me shake my head. So out of all the other people in the "race", who all want to win - you have to be the one that does? What pressure does that put on you? And if you don't win, does that mean you're a loser, and that everyone else who didn't win is a loser as well? That would mean that one person feels great and everyone else feels lousy - again and again and again in life.

I don't know how you feel about that, but I personally think it's pretty flawed.

What about us females - what have we been fed? "Be a lady!", "Good girls don't do that!", "What will the neighbors think?", "That's not a job for a woman!" Wow. What the heck? What is a "lady?" Guess what, we all fart and burp and poop and that doesn't make us any less womanly. In fact, we're in trouble if we don't have these normal bodily functions. If "good girls don't do that" and we do these things, are we now set to believe that we are in effect "bad?" Then we get the gift of being bad for the rest of our lives? As for the neighbors, guess what - to a large degree they're dealing with their own issues and if they care about ours then they need to find other things to do. Newsflash - it's not healthy nor enjoyable to live your life for the neighbors. And they don't really care that much anyway. "That's not a job for a woman" just took away your power, didn't it? All of a sudden there's a huge list of things that we're not meant to do simply because we were born in to a body that didn't have male parts. Does that make sense to you?

What else is in your box? Consider the walls and barriers that have been placed around you by others and even by you. Is it in your box that nobody in your family has ever graduated so neither will you? That your parents are teachers or doctors and it's expected that you will follow in their footsteps? How do you truly feel about the religious, gender, sexual identity, cultural, educational, relational, health-based or familial expectations and walls that you've accepted your whole life. What about the stuff that society has placed in our boxes? If you are a part of society, then don't you get your own say on societal beliefs? Honestly, take a mindful moment and ask yourself how you *really feel* about the box you're in. It'll take a minute. I'll wait.

How many of us have been hurt emotionally, by ourselves and by others for not living within the confines of our box?

A large part of being mindful is understanding who we are, including the self-limiting beliefs that bind us. Become familiar and understand what's in the box that you surround yourself with. Then give yourself permission to step out of it and abandon it, because that's what you deserve. **Eliminate the box**

Shine your light

"What lies behind us and what lies before us are tiny matters, compared to what lies within us." - Ralph Waldo Emerson

You are special and you are unique. You are amazing. You have so many gifts and talents that you can't even imagine them and it is now time to honor that truth. It doesn't matter what you believe or what you have been told in the past - the only truth is that you have incredible gifts to share with this world.

Take some time to honor the beauty in you. What do you have to share with others that makes you feel good? Perhaps you have works of art inside of you just waiting to come out and inspire people with. Maybe you have a sense of humor that can make even the darkest of times a little easier for someone. Does your soul light up when you spark curiosity and wonder in children?

Your shine is inside of you for you to see and it's inside of you for everyone else to experience as well.

Earlier this year during the summer I rounded the corner to our house to find the contents of our garbage can strewn about the cul-de-sac. Although we all try to time putting out our cans in order to limit bear snacks, I had to leave early for work that day and didn't think I'd be home in time to beat the truck. I groaned to myself, fast-forwarding by one minute to when I'd be picking up trash in my dress, heels and hosiery as fast as I could before the garbage truck arrived. Sure enough, the truck followed right behind me and I apologized for the mess and told the lovely lady that I would have it cleaned up in a moment. She was shocked and insisted that that's what her gloves were for and that she would do it. Then the gentleman driving the truck also hopped up and started collecting all of the garbage in the lid and bringing it back to the truck. I explained that we have trouble with bears in the area and they assured me that it happens all the time. The three of us quickly cleaned up the mess - they were friendly, professional and cheery. They made my day through their beautiful and helpful attitudes. Although we have competition for Sanitation Providers in our city, I am forever loyal to the one we have. Every single week no matter who is in the truck, they just shine. Being just one of their thousands of clients, it makes me smile to think of how they change the world by just helping others and enjoying their job.

Shining is not about ego. It is not about being "better" than anyone else at what you do, more so it is about being courageous enough to be the best you that you can be. It is about being so grounded in putting out your authentic goodness that just being near you inspires others to do the same.

Shine your light

Love from afar

"Since love grows within you, so beauty grows. For love is the beauty of the soul. " - Saint Augustine

All of us have love to give and to share - it's in endless supply inside of us just ready to be tapped at any moment. There are times, however that sharing our love in person is not a viable option. Whether the person or object is with you or not should never stop you from sending them love. At any time that you cannot be with whom you would like to give love, then offer it from afar.

Sometimes there may be a person or people in your life that are struggling and being near them in their current state would be unhealthy for you and/or for them. Recognizing this may be easy or it may be excruciatingly painful, and either way you should be proud for acknowledging and accepting the reality of the situation. When this is the case, send them love from afar. If you are upset about something that you have read in the paper or seen on the internet or the news and you cannot be there to help in person, love from afar. If the relationship is toxic for either of you then don't put anyone in harm's way, and simply love from afar. Love is a form of energy and energy moves. Only good can ever come by sending someone or something love.

Your love makes a difference in this world every single time you feel and experience it. It makes a difference to you, it makes a difference to those around you, and I invite you to trust that it also makes a difference to those that are *not* around you. Love from afar

Randomly offer kindness

"Service to others is the rent you pay for your room here on earth." - Muhammad Ali

Each and every one of us is a never-ending container of possible gifts for this planet and for those who inhabit it. It can literally take one second to give an offer of kindness that creates positive life-altering events.

Take a moment to reflect on something nice that a stranger has done for you. That adult that gave you the few cents you were short when you were a kid at the candy store, that person who held the door for you or helped you to pick up what you had dropped. That anonymous human being that turned in your wallet including all of its contents to the lost and found. Maybe the act made you smile, maybe it made your day and maybe it even shaped your life. It's astounding the impact a small gesture can have.

Imagine how different this world would be if more people mindfully looked for opportunities to randomly offer kindness to one another. This could be done in place of judging one another, as it takes the same amount of time but offers way more health benefits for everyone involved. Is that a world that you would want for your loved ones and for yourself? The possibility lies with you and with me - I'm up for the challenge to make a difference, are you?

When you look for opportunities to randomly offer kindness, you'll find them everywhere every single day! From the moment we get up in the morning until the moment we snuggle in to bed we can actively pursue the act of giving the gift of kindness.

If we are mindful, we will find more options to offer being kind than we ever could have conceived. Offering to reach something for someone in a store, holding the door at the drycleaners, giving someone in the lineup behind you your spot if you know your transaction will be long, letting a car in during rush-hour traffic or helping someone who looks confused or lost can all make a difference. Flash a heartwarming smile or offer words of encouragement to a child. You could even offer kindness that doesn't need an instant recipient such as picking up a piece of litter and putting it in the trash or returning an abandoned buggy to the store so that it doesn't roll and scratch someone's car. Any of these things can start the change that we would like to see in this world. If the planet was a bank account and every single act of kindness was a deposit, just think of how rich life could be for everyone with just one or two kind acts a day. **Randomly offer kindness**

Through not for

"You are not called to be a canary in a cage. You are called to be an eagle, and to fly sun to sun, over continents." - Henry Ward Beecher

So many of our parents feel or wish that we were born for them and should behave as they would like us to behave. So many of us parents feel or wish that our children were born for us and that they would behave as we want. We need to be mindful of the fact that everyone, including us, is born "through" someone, but never "for" someone.

During a session with one of my clients several years ago, Dianne confided in me that she had her one child, a son so that she had somebody to take care of her when she became too old to take care of herself. She told me that the marriage with her second husband was not going well and that she was comforted in knowing that she would move in with her son, his wife and their children when the time came. When I asked her how her daughter-in-law felt about that she replied, "She doesn't know yet. Actually, neither does my son. But that's why I had him." At first, I thought she was kidding however her deadpan facial expression told me that she was completely serious. I then asked, "What if they don't want you to move in with them?", to which she reiterated to me again, "That was the entire purpose I had him."

She went on to tell me that she didn't even want or particularly enjoy kids but that she knew she wouldn't want to be alone when she was old so hence, she had a child.

I wondered about her son. What would he say when the day came that she informed him that she was moving in with him and his wife? Did he know that he was solely born to be her caregiver? If he did, how did that make him feel? If he didn't, how would he feel if she ever told him? Would he invite his Mother to live with him and his family in his house with open arms? Would he do it begrudgingly? What if it impacted his marriage in a negative manner? What would happen if he told his Mother, "no"?

So many times, people shared with me that they entered into the profession that they're in, married the person they married or had children simply because that's what their parents wanted. There's no judgement in that, just a question - are you happy?

Being mindful includes being curious about who we are and to question our thought processes and beliefs. So many of the people I worked with had grown up thinking that their lot in life was to please their parents.

Most of us want our parents to be happy and even to be proud of us. Most of us want our children to be happy and successful. When people love one another unconditionally then that should come naturally, without expectation.

Give yourself permission to mindfully be curious around your own feelings and beliefs with your parents, caregivers and children. ***Through not for***

I've got me

"We spend our lives avoiding the situations that help us grow. It's when we stay with uncertainty and discomfort without trying to fix it that we connect with our own innate joy, wisdom, and love." - Pema Chodron

Wouldn't you love it if you were guaranteed at birth that there is one person that will be with you in your life no matter what? This person will be with you during your moments of highest exaltation and times of deepest despair. Through every change in life, every small or momentous event - right there. No matter what you ever say or do, no matter what emotional rollercoaster you are riding, they will be with you. Wouldn't that be amazing and comforting? That person exists of course, and it is you.

Thank goodness you've got a built-in best friend for life. How fortunate you are!

You have an opportunity for your entire life to be the friend to yourself that you always wanted to have. You can give yourself a hug in the comfort of your bedroom or a bathroom stall if you're having a bad day. You can remind yourself that everyone gets pimples when you wake up and see that big zit on your face right before you go in to your important meeting. You can smile at yourself in the mirror and tell yourself to "knock 'em dead!" today. You can take a deep breath and enjoy every single second with a loved one and you can take a deep breath and calm yourself down after a harsh word to yourself or someone else. You can tell yourself that you're going to make the next hoop, the next mile or the next promotion. You can be that friend that reminds yourself when a good show or a little nap is exactly what you need.

If you notice that you don't talk to yourself the way you'd like someone to speak to you for the rest of your life, that's OK! Just change it. You have a lifetime to master how you care for, and comfort yourself and you get to start right this moment.

No matter what happens, you've got you. Remember when you're afraid, "I've got me." Say to yourself when you've knocked it out of the park, "I've got me." When you're feeling lonely or sad, "I've got me." And when you're having so much fun, "I've got me." You're so lucky to have you! Nobody else in this life will ever have you the way that you get to have you. Make it a relationship that you love having, and then remember that you have it.

<u>I've got me</u>

Something that happened

"There are far, far better things ahead than any we leave behind." - C. S. Lewis

Many of us get caught up in certain events that happen throughout our day, whether they be "good" or "bad." For those of us who have ever been in school or who know someone who has been in school, a common comment may be, "I had the worst day today - I failed my math exam." Let's use that example and take a deeper look; is there possibly anything else that happened during the day, or did that person just wake up and get transported to the exact moment where they found out they failed their math exam and then get zapped back home to tell you it was the "worst day"?

There were a number of other things that likely happened in this story; that person woke up in a nice warm bed hopefully after a restful and restorative slumber. Maybe they had a nourishing breakfast and either drank and/or cleaned themselves with clean water that came directly from the tap. They possibly had the physical ability to walk to school or had someone give them a lift in a reliable car. They got to go to school and be taught, clothes on their body, pen and binder in their locker and they got to interact with others before they came home to a roof over their head with someone to talk about the math test event. Is that really the worst day? That failed math test was just "something that happened" in their day, but it wasn't their day.

We learn early in life that it's socially acceptable to talk about and revitalize the "something" that has happened in our day, our week or even our life and that we will likely be able to find a compassionate ear to listen to our complaint. What we don't realize is that this process is programming our brains to focus on the negative and allowing us to completely skip the positive things that have happened before and after the "something."

Imagine how different our lives would be if we took a look at every single "bad" thing that ever happened to us and saw it just as that "something that happened." That moment or event is gone now - finished, completed, ended, over-and-done with. There were a lot of other somethings that were really good, great even - and hopefully they far outnumber the other stuff. We can move on - because that "something" is completed and we're stronger now for living through it. Maybe that student got help after the exam and now they know that they just misunderstood the questions. Perhaps that failed test will spur them to action and they will get tutoring or help from a teacher or peer, and it may one day inspire them to become a tutor themselves. Hopefully they'll learn that it was just a moment in time, that it has passed now and that they can make decisions based on that "something" that will help them in the future.

Or they can let it ruin their day, week, year, life - the choice is up to them. The choice is up to us.

I would never discount what has happened to people and some of the tremendously painful and negative experiences that we go through in this life. My suggestion is that you try to view it as a "something." It doesn't define your day or your life and it can make you stronger. It doesn't have to define you if you don't allow it to.

There's a lot to be said about having professional support with these "somethings" that occur and I definitely recommend it if this is what can help you.

One of the many things that I love about "something that happened" is that I have learned to project to the future when I'm in the middle of a something to when the something will be in the past. For example, when I was 9 months pregnant with a flat tire in an uncomfortable part of town needing badly to urinate with no bathroom in sight, it was actually comforting to know that as soon as this issue was resolved that it would soon become a something. Then instead of allowing it to ruin my day, I can one day write it in a book and laugh (or grimace) and be glad that it's over. That fact alone helps me to calm down and do what needs to be done in the moments of the "something" as quickly and efficiently as I can so that my "something" becomes a "something that happened." Give it a try. **Something that happened**

Slow things down

"No one can get inner peace by pouncing on it." - Harry Emerson Fosdick

Although our egos want to convince us otherwise, there is scientific data that supports the fact that multitasking makes us less efficient. Well, doesn't that just change things....

A client of mine once admitted that he used to bob and weave through traffic trying to get everywhere he could as fast as humanly possible. He had convinced himself that he was saving tons of time with his zipping in and out of the slow cars that surrounded him. One day he had been doing his normal routine in rush hour traffic and after a considerable amount of time he looked over to see a car that he had passed long ago still in the slow lane right beside him. He realized that his tactics hadn't saved him even a minute but had put him and others around him at risk.

He bravely admitted that when he started driving in the slow lane to work, he noticed that he was arriving only a couple or a few minutes later than the times when he drove dangerously. He said it was a huge eye-opener for him and he learned that it makes more sense to slow his driving down than to fight his way through every day.

This is a good learning lesson for all of us in a variety of different ways. Not only asking ourselves if we really need to drive so fast but also if we need to respond to that text immediately or even answer the phone simply because it's ringing. It may be more responsible for us to wait until we're in a good head-space before we engage with others in any type of communication. Does rushing through the mall really get you to the store that much faster? Does running around at home being distracted by every single thing that needs to be done really make you more efficient? Even if it did (which it likely doesn't), how does it make you feel?

There's a big difference between doing things slowly and slowing things down. As with most things they both have their pro's and con's. Being mindful allows you to discover what they are and what works best for you.

Make a conscious effort to give it a try; you might like it. Slow things down

Pluck the weeds

"Nothing will work unless you do." - Maya Angelou

Mindfulness takes effort and work, although it's meant to occur gently and tenderly much like someone might carefully tend to their garden. Just like in any garden we never plant weeds but somehow, they pop up anyway and then need to be addressed. Sometimes when we are in all of our fervor planting away, we don't see that a small weed has germinated. Negative thoughts can be like that as well - they start small but if left unattended they can grow until they become unruly. There's a reason for the saying, "growing like a weed."

Mindfully realizing that you've got a weed in your mind is the first step. It's that mean comment, either about yourself or others, that desire to compete unnecessarily, the urge to point out that someone else is wrong and that you're right or any other list of negative thoughts and emotions. OK, so you've recognized it's there - great job! Now it's time for step number two to simply pluck it.

This entails changing that negative thought to something positive and you can do it in a variety of ways. Maybe you like oppositional thinking where you turn the comment around in your head to something constructive and useful. Perhaps you may find it helpful to think something nice about the person, place or event instead. I've used lots of methods and my favorite technique as of late is to just allow it to drop like a lead balloon so that I can step over it and get back to doing something more productive with my thoughts. Plucking weeds is reminding yourself that you don't want negative thoughts germinating in your garden.

I'm not saying that weeds aren't going happen. I can't think of any garden in the world that doesn't need to be tended to in order to have the weeds removed. It's natural and not at all a reason to get upset about it. Getting upset about negative-thought weeds just causes more negative-thought weeds. Be forgiving with yourself for the thought and then pluck it using your favorite technique.

Just as having the occasional weed doesn't mean that you're a terrible gardener, having the occasional negative thought doesn't mean you're a bad person. I've been practicing mindfulness for years and still have to pluck weeds when they pop up.

One of the coolest things about this practice is that the more negative-thought weeds that you get rid of, the better you get at finding them as soon as they appear. The benefit is that each one you get rid of makes you better at tending the magnificent estate of your mind. Happy gardening! Pluck the weeds

Find your path

"If you want happiness for an hour - take a nap. If you want happiness for a day - go fishing. If you want happiness for a month - get married. If you want happiness for a year - inherit a fortune. If you want happiness for a lifetime - help others." - Chinese proverb

There is no trickery in mindfulness. There are no undercurrents or mind-games. Mindfulness is about finding yourself, not about fooling yourself.

We all have a path to happiness and fulfilment. Every single one of us has the opportunity to find and live a blissful life. We find our path by closing our eyes and looking within versus what many of us try to do, which is frantically searching outside of us. When we close our eyes, we can strengthen our skill at listening to our bodies versus choosing to listen to the craziness and inaccuracies of our thoughts.

Here's a question with no strings attached - it's a quick, simple and private "yes" or "no." Are you happy; *truly* happy?

If you are, then that's awesome and keep going. If the answer is no, that's awesome too because the first step in finding your path is realizing that you're going the wrong way.

The second step is following what does make you happy. What's the most nourishing decision that you can make for yourself right now? What can you do differently to bring more happiness in to your life?

Countless clients have told me that they became "stuck" in their career due to the pay, the promotion, the proximity to home and a whole list of other reasons. Sometimes a short-term job turns in to a career and then we're left wondering how it happened and how to get back on track to following our dreams or callings. You're not alone.

For others, they are already in the profession that they love. Being great at what you do and passionate about doing it makes a difference in this world every single day. Your efficiency, your knowledge, your attention to detail or your creativity helps people to do what they need to get done.

Please actually give this some thought, and I don't mean while you're washing the dishes or during the commercial break of your favorite TV show. Respect yourself and your happiness enough to give this some deep, meaningful thought. What would you do if you were moving towards happiness?

My personal journey has already had a million steps and I continue to travel to my path, although I'm closer now than I've ever been. It's taken years to be right here, right now writing this book and has included things such as taking up meditation and reading books that move and inspire me to leaving an entire career behind me on the faith that starting something entirely new is what I'm meant to do. Life is simple, but it's not always easy.

It's quite possible that you already know the steps that you need to take to lead you closer to your path or further along it. Follow the voice of your heart and not the voice in your head. Getting closer to your path feels good and getting further away from your path feels the opposite of good. It's literally that straightforward.

A wonderful place to start can be through the act of giving. You may find that it feels inspiring to donate your time to a charity or good cause. You could also just try something new that you've always wanted to. Maybe the reason you've always wanted to try it is because it's a part of your path. Go to a bookstore and find a book that just seems to call out to you, even if it's completely different than anything you've ever read before. Talk to others about what interests them and see if you learn something new or if it piques your curiosity. Take a class somewhere. Even if you don't end up liking it then you've learned something new about yourself! When you start to feel excited and passionate about something it can be an indicator that you're on to something. Remember that life is a journey, not a destination. **Find your path**

Simply a container

"In the sky, there is no distinction of east and west; people create distinctions out of their own minds and then believe them to be true." - Buddha

This temple called a body is the container that you were placed in when you were born and it is yours to keep for the rest of your life. Wow! What a huge gift and a big responsibility.

So cool that all of our containers are incredibly different. Even if you're a twin, the container you have is still distinct from your brother or sister's. How remarkable is that - twins have the same mold but they *still* come out unique. It's actually astounding if you really think about it.

Something else that is astounding about these containers is that many of us have been brought up to believe that the containers we live in somehow exhibit the amount of worth that we should put on ourselves and other beings. If the container is considered by others to be attractive then we'd rather be seen hanging around with that container. If the container has lots of money in its pockets then it seems to sometimes be allotted more in life. If the container has a job with power then sometimes, we respect that container more than others. In too many tragic cases containers have been viewed by some as "less important" simply because of the way they look including the color that covers them.

We have no say in this life regarding the container we were born in to. I was born in to a container that is now six feet tall. It was very difficult being in a container that was so much bigger than everyone else's when I was growing up. For many years I thought that the only way to feel more confident about my large container as a female would be to have a male partner whose container was taller than mine. Thank goodness I got over my ignorance about containers because my life would have so much less depth and meaning than it does now. My husband's container is 5 feet, 2 and 3/4 inches tall. When I am in heels, I am more than a foot taller than my male partner. It is funny to watch some people's reactions when they see us together because they think we make a ridiculous couple simply because the female container is significantly taller than the male container. If either one of us cared what some people in society thought, we'd be missing out on the most wondrous relationship that either one of us could have ever fathomed.

One of my clients who happens to be in an Indian container is single and has said that she will only marry a partner who is also Indian. When asked if she believes in soul mates, she responded in the affirmative so I pressed and said, "What if your soul mate is Chinese?" Mandy informed me that it meant that they would not be together in this life time. Is her soul mate any less lovable just because they're not in the "right" container?

Whenever I look at anyone now, I try to remember that it is simply their container. Some containers are homeless and some drive nice cars. Some containers are decorated and other containers have various marks on them. Some containers have fancy hair and some have no hair. Some containers speak articulately and some containers are not capable of saying a word. It doesn't mean that they don't have anything to say, it's just that their container cannot speak.

All of these containers have a soul living inside of them. They all house a person. That person matters no matter what their container looks like, sounds like or owns. Every single soul inside of every single container is worthy of respect. Mindfulness includes seeing past the container; including your own. You may be delighted by what you find inside. ***Simply a container***

Strong versus skinny

"Goals must never be from your ego, but problems that cry for a solution." - Robert H. Schuller

These three words, like many other phrases in this book, shifted my world and helped to promote my growth of a healthy body image. Being truly mindful - compassionate - about our body image is one of the components I find that people struggle with the most and I was no exception.

As a child I was overweight and at times ridiculed quite intensely for it. On one memorable day one of my brother's friends informed me that I was "like a cow because I had five stomachs." You get the point.

Like many others on this planet, I really struggled with my body - in fact, I struggled *against* my body. Although I battled and got my weight to an "acceptable" level I was still incredibly self-conscious about the spare tire around my mid section. Going swimwear shopping was my own personal hell.

In 2013 I realized that a number of things in my life needed to be addressed and I started once again on my path to weight loss.

In an effort to really "support" my weight loss program, I went out and bought a tight little work out outfit consisting of cute shorts and a bra top that were several sizes too small for me. I figured that by being humiliated, shamed and disgusted every single time I passed the hall closet mirror to head to my elliptical trainer that surely *this* time it would work.

As you likely guessed, it didn't.

What *did* change everything was a work-out tank top several sizes too big found at a local retailer for a whopping $12.99. The shirt stated in large letters on the front "STRONG is the new skinny." I stood there at the store. Stunned. Healthy finally made sense to me.

Thank goodness for that shirt.

I started eating more salads and choosing extra-lean ground beef not because I was on a "diet" but because it was going to make my body strong. I started working out harder, choosing exercise that I actually like to do, and having more fun with it because it wasn't about weight-loss anymore; it was about making my muscles stronger. My body responded almost immediately and with ceasing to weigh myself on a daily basis, I could start to *feel* the difference in my energy, muscles and shape versus looking for the change on the scale.

There's a gruelling hike about 90 minutes from my house and I started doing it on a regular basis with a completion-time goal in mind. I haven't hit it yet; maybe I will in this decade, maybe in the next one or maybe never - it doesn't matter. What matters is that I'm having fun and I'm proud of myself. I wear my oversized tank top every time. One day at the top a woman with a bunch of her girlfriends approached me and said it was the best shirt she'd ever seen in her life. I agree with her. (As a side note, it's possible that she originally approached me due to my heavy breathing, red face and profuse sweating - that lovely concerned citizen was worried I was about to have a heart attack. But that's beside the point, she liked the shirt.)

Gone are the days that I "punish" myself for being larger than I want to be - I'm strong and healthy and that's all that matters. Isn't it time for all of us to love every single inch of ourselves and focus on what's reflecting from the inside versus what's reflected in the mirror or on the scale? **Strong versus skinny**

Light will follow

"The most beautiful people we have known are those who have known defeat, known suffering, known struggle, known loss, and have found their way out of the depths. These persons have an understanding of life that fills them with compassion, gentleness, and a deep loving concern. Beautiful people do no just happen." - Elisabeth Kubler-Ross

There are many sayings that elude to the light shining through after the dark. They include, "Without the dark, you wouldn't appreciate the light" or, "You can't see the stars without the dark" and they all equate to the same thing; *the bad times help us to truly appreciate and savour the good.*

I know that can be hard to take, believe, or subscribe to when you're in the middle of a black-out with no end in sight, but the moment of change always comes. Perhaps it just hasn't come yet.

Regrettably, as many people do, one of my best friends went through a very acrimonious divorce. The time ticked on and the costs continued to escalate for all parties. It took a lot of energy and caused a lot of emotional hardship for everyone involved. During that time, it was difficult to imagine that there would ever be an end. As another New Year rolled by sometimes things seemed hopeless.

Knowing that sometimes we need the dark side to really appreciate the good side helped him to get through. He was confident that when it was finally over everyone could move on in life, to something that was exponentially better than before. He learned to fully appreciate the dark times, knowing that when the dark lifted and the light eventually arrived that it would feel amazing. And it does.

We've all been through, are going through, or will possibly one day go through a dark time. Perhaps that darkness will be regarding the emotional or physical health of yourself or someone you care about, maybe it includes family or friends, an unhealthy relationship, finances, safety, housing, bullying, loss of a job or any other difficult time that we humans go through.

These three words are something that can make those dark times a little less dark; remembering that there is always a light at the end of the tunnel. There is a well-known joke that maybe the light at the end of the tunnel is a train coming toward you. If that happens to be the case, you're now provided with another opportunity to buckle up for the ride and have confidence that clearly there is an end to the tunnel - the train just came from there!

Take care of yourself during these dark times and be as kind as you can. Take a deep breath, roll up your sleeves and get done what needs to be done - one moment at a time. Be gentle and loving to yourself, remember that when we forgive others, we release ourselves from prison as well. Love yourself, encourage yourself and be your own very best friend. Have faith; the stars will twinkle and the sun will shine. Keep moving with hope and faith that things will get better. **Light will follow**

Relax with nature

"The best way to capture moments is to pay attention. This is how we cultivate mindfulness." -Jon Kabat-Zinn

One of the most important aspects of our health is the state of our nervous system. A calm, relaxed and stable nervous system will help us with everything from our mental health (including the ability to respond positively to unexpected events) to our physical health (fighting off the common cold or even battling with a disease).

A perfect way to help affirm a resilient nervous system is to be with nature. This can come in many forms depending on what works best for you.

If you're someone who leans towards auditory stimulation then listening to nature sounds will help to soothe you. These sounds may include the wind through the trees, listening to birds singing, the rain coming down, water babbling in a brook or stream, the ocean waves or any other host of natural phenomenon.

Many people prefer visual stimulation so sitting comfortably in a park looking at flowers, lying down in your back yard watching the clouds or the stars, going for a walk, run or hike and seeing all the trees or watching water may be just what the doctor ordered for you.

If you find touch nurturing for you then laying on the grass is perfect. There are many plants that are safe to touch and feeling their softness, smoothness, delicateness, coarseness or any other feeling sensation is perfect. Spending some quality time snuggling with your pet is also a great way to soothe yourself.

I really enjoy the beautiful smells of nature so being out for a walk or sitting in nature gives me an opportunity to smell the way that plants respond to the sunshine or rain. The scents of flowers and freshly cut grass are amazing, as is the salty ocean air. A good friend of mine is embarrassed to share with people that she absolutely loves the smell of cow manure. She said that she doesn't admit it to anyone, but that the farm smell is as close to nature as she feels anything can be.

Lastly of course are those of us who find the taste of nature healing. This is where you take your time and truly slow down to fully taste healthy food. I've been told that there is no better taste than a carrot picked straight from the garden. As a child we used to grow peas in our back yard and there were never any available for dinner because we kids had already eaten them.

Many of these things are also available for further integration through use of a mediation and you can find information all over the internet or in meditation books or classes. There are walking meditations that involve slowly taking one step at a time with absolutely no concern about how fast you get there; it is about taking the step, feeling the ground under your feet, your muscles supporting your body and just being part of the outdoors. I enjoy walking meditations where with each step you stop and think the phrase, "I am here."

Eating meditations are different as well and they can involve taking a small portion of food, such as a raisin, a raspberry, a walnut or any other healthy food and fully being present for every sensation of eating. You can quietly smell the item, take a small bite and be immersed in how it feels in your mouth. Eating a raspberry one tiny little piece at a time takes a lot longer than you would think. (Sometimes when I eat my chocolate covered jujubes, I'm pretty sure I could enter a meditative state; however, I wouldn't say that counts as getting closer to nature...)

My point here is that the benefits of calming your nervous system through the enjoyment of nature is a great way to make you feel more grounded. Even five minutes is better than nothing and will make a difference if you take the time to slow down, breathe deeply and enjoy. Relax with nature

Best I can

"Serenity is knowing that your worst shot is still pretty good" - Johnny Miller

This is a phrase that I started using professionally and found that it could be equated to absolutely everything in life. It took me a long time to get to this point because I always wanted to be perfect. I was proud of the fact that I consistently responded to corporate emails within a few hours whether I was at work or on vacation, in the office or out in all-day meetings - it didn't matter. I gave everything I had to each presentation I did, every meeting I led or participated in, and all of my client interactions. It got to the point where I was so out of control that I could no longer sleep at night and would often sneak out of the bedroom at two o'clock in the morning, take my computer offline and respond to emails that I would send out later that day. On the rare times that I would forget to take my computer offline I would receive an email from someone saying, "What the heck were you doing working at 3:24 in the morning?!"

Eventually, being a workaholic no longer worked for me. I did not want to be controlled by the drive inside of me that was doing a hit-and-run with my life. As hard as it was, I had to realize that sometimes all I could do was the "Best I can" and that would be it. Those days of document-writing in between feverish spells from bronchitis while lying in bed were gone and I was never going to return.

Although it was incredibly scary to let go of being "perfect" it was also overwhelmingly freeing. I felt like I could breathe again, like I was allowed to be human and I was allowed to be imperfect.

Don't get me wrong, I still worked hard to keep my employer and my clients happy and my reputation still mattered to me. The difference was that I no longer worked *past* the one hundred percent mark. That means that on the days I didn't feel well, I allowed myself to take it a bit easier and not push so hard. The work still got done - it just maybe didn't get done that day and for the first time in my life I was OK with that.

That mentality of "Best I can" easily morphed in to my personal life as well. Being a caregiver, whether it is to children or adults can be incredibly taxing and sometimes we feel like we have to just keep going and give everyone everything we have. The thing about giving everyone everything we have is that then we have nothing left for ourselves. That's a very dangerous and precarious position both for us and for the people we're caring for.

Yes, at times I was still Super-Mom and did all the cleaning and helped the kids with their homework and played games with them or took them to the park. And sometimes I would just do the "Best I can" and end it at that. The kids were surprisingly amenable to me putting in more boundaries for myself and learned to play more on their own, with each other or with their friends. And there were times when looking at the call display and deciding *not* to answer the phone when I didn't have it in me that day felt really liberating.

Give it a try. **Best I can**

Stop and Breathe

"I am indeed a king because I know how to rule myself." - Pietro Aretino

Simply because we are human, we have the right to make mistakes. Auspiciously due to our humanness, we have the capacity to learn from our mistakes and adapt so that we can elect to make different decisions in the future. The complex skill of reflection and environmental adaptation is a unique and exceptional gift.

Actively pursuing the opportunity to "Stop and breathe" is a nurturing and responsible decision. It often buys us the time we need to get in touch with our true intentions and to make better decisions. Many times, it may save us the need to reflect in the future on how we could have done things differently because we would have made the best decision at that time through successfully using the technique.

Jon Kabat-Zinn, who is an expert in the field of Mindfulness is also the brain-child behind the mindful acronym STOP. It represents the following:

Stop

Take a breath (or multiple breaths, if needed)

Observe what is happening (around you, inside of your body and mind)

Proceed with mindfulness

This technique alone has the power to change the world. Clearer heads would prevail and better decisions would be made. Relationships would be preserved, feelings would be protected and humanity's true nature would have the opening to shine through in all of our actions.

The first part of this technique is to stop yourself. When you feel anger arising, stop. When you're about to make a reckless and impulsive decision, stop. If you're about to put something in your mouth that you'll regret later or about to make a choice that will likely lead to disappointment, stop. In order to change the cycle, we must first interrupt the process.

This is easier said than done and becoming in touch with the sensations in your body is the best way to decipher when it's time to stop. For me I feel anger as a tightness in my chest, a fire that wants to explode outwards. As soon as I feel that well up inside of me and I'm about to blow, I stop. Impulsive decisions come to me through a rapid heartbeat and an intense desire to act immediately. There is a thought in my head that if I don't act on the thought instantly that I will lose my opportunity. I know with certainty that whenever my heart pounds like that and the thoughts are screaming at me to say or do this "brilliant" thing, that I will regret it very shortly after. It's so difficult to tell that instinct to stop because it fights back like a lion that refuses to go in to its cage. When that happens then I know that it *really* isn't a good idea.

When the word "Stop" screams in my head, I listen. Whatever I'm saying, I stop. Whatever I'm doing, I stop. It's not easy and it doesn't happen without practice. I've missed the mark a million times by screaming "Stop" in my head and then continuing anyway only to later regret my initial decision to override the message. If you need permission from someone other than yourself to be OK with messing up sometimes, you'll find it in the first sentence of this chapter. We humans mess up. We humans get the opportunity to try to do better the next time. Life is a journey and sometimes the journey involves door dings and flat tires along the way.

The second part of "Stop and breathe" is to breathe. Yep, take a breath and truly *feel* it. *Follow* your breath with intention; feel the air in your lungs, the rise of your chest and shoulders or the sensations in your nostrils. However you are aware that you're breathing, be aware that you're breathing. Allow yourself to be fully immersed in your breath, as though you are getting lost in it for just a moment. You'll know when your pre-frontal cortex has come back on-line and the ability to make more rational decisions has returned. It's a good feeling. For me, it physically feels as though I've returned back to my body and I'm in the driver's seat again. I can feel when I'm back to being "me."

After the breath (or two, or three...), observe what is happening all around you. Be aware of the thoughts in your head, be inquisitive about the sensations in your body and be curious about how the other person or people are doing. Observe the event as though you are an alien that has just been transported in to that container and you're so interested to know what the human experience is like. Be a scientist doing research in your own body.

When you're back in to your full capacities then proceed with care and mindfulness. It may be that you're able to make the most nourishing decision right there, or you may find that the most nourishing decision is to acknowledge that you're not in the best position to make a decision at this time. Both of those options are a huge win. They both indicate that you stopped the cycle. Way to go!

Even if you don't get a hold of yourself with the "Stop and breathe" technique and you end up not stopping and not focusing on your breath but decide afterwards that you really should have, that's still more of a win than if you hadn't come to that reflection at all. It indicates that you're one step closer to being able to do it successfully - that step is thinking about it.

There are a number of times when this phrase would not be suitable, including if you're being chased by a bear or trying to escape a building fire. However, there are just as many if not more times that it would indeed be suitable. In fact, there are a multitude of times when this practice would be prudent. Stop and breathe.

Reduce the caffeine

"Common Sense is that which judges the things given to it by the other senses." - Leonardo da Vinci

You don't need a health lecture from me and I'm not going to give you one. This phrase is just meant to be a reminder that if we can cut down on the caffeine even a little bit that you may find it makes a significantly positive difference.

I adore my Earl Grey tea and always have, literally having been brought up drinking it since before I could read (yes, the caffeinated version). I'd read the articles in the paper, the tips in health books and heard the lectures from professionals but nothing was going to stop me from having my daily pot of tea.

As I got older and more in tune with my body I started noticing certain things like the fact that all of a sudden, I was in deep need of eye drops due to the lack of moisture in my eyes. In addition, I sat down one day to do a meditation and when I closed my eyes, I thought I was having a panic attack. That would have been odd for me as I don't normally suffer from panic attacks so as I silently investigated what was happening in my body, I realized that my pounding heart and sweaty underarms were due to the sinfully delicious, large-size chai latte I had just consumed. That got my attention.

There was an article that came out on-line in the first quarter of 2017 by Dr. Travis Bradberry. Two of the things I liked most about it was that it was short and to-the-point. My thought is that if you're going to beat me up about one of my vices, at least make it quick. I won't tire you with too many of the details but along with his statements that it has an impact on your emotions, creating anxiety and irritability, as well as gets your adrenaline pumping and affects your sleep he also included a picture of the body with little offshoots of how it impacts us physically. It was noteworthy to me that all of a sudden, my muscle twitching and sensitivity to touch made sense. The article also explained my issue with dehydration. Abdominal pain was listed there too and I thought of all the clients I'd had over the years who told me that they had chronic stomach aches. I couldn't help but wonder how many of them may have been suffering due to possibly drinking too much coffee.

Don't get me wrong, even my Doctor of Traditional Chinese Medicine has sheepishly admitted to me that she loves a cup or two of coffee in the morning so it's not necessarily something that people want to give up wholly. Hey, what's better than a crisp Coke on a hot day or Cream of Earl Grey on a weekend morning? (OK, a few things but you know what I mean...) The point is what we've all heard from Doctors all over the world so many times; moderation. **Reduce the caffeine**

A visual reminder

"Time is a very precious gift - so precious that it is only given to us moment by moment." - Amelia Barr

We need to take advantage of every moment in this life and sometimes we need a little help - that's where a visual reminder can be a game-changer.

Years ago, I was conducting a joint coaching session with one of my employees at the Bank. As her Manager one of my responsibilities was to silently observe her during client interactions and then provide her with feedback.

Jenny was a positive person with a great sense of humor and a smile that could light up a city block. I was shocked when I noticed that she would smile as she called her clients over to the wicket and then do her best Grumpy-cat impersonation for the remainder of the interaction. Needless to say, this was the "Action Item" that we would focus on for our session. Jenny was surprised to hear that her smile disappeared as soon as she was face-to-face with a client and when asked for a suggestion as to what she could do to help her remember to smile she was at a loss.

When asked simply, "What makes you smile?", she got embarrassed and replied, "In life?", to which I responded in the affirmative. She blushed a little and said. "my dogs." Her homework for the day was to find a picture of her dogs small enough to fit on a corner of her computer screen so that she could post it as a reminder of what makes her want to smile.

The next day after serving several clients Jenny came up to my desk with that huge grin on her face and said, "Thanks for the suggestion to put my dogs up there - I honestly can't stop smiling every time I look at them!"

That's all it took. It didn't need to be a big deal. Sometimes we all just need a reminder of what it is that we'd like to focus on. In high school I had a girlfriend that had a red dot-sticker on her watch and when I asked her what it was for, she said it was a suggestion from her therapist that every time she saw the red sticker, she would take a deep breath to help reduce her anxiety levels. The anxiety caused her to look at her watch a lot through the day and there it was every time - a reminder to slow down and take a breath. Being a young teenager, I asked her dubiously if it actually worked (you know - those adults don't know *anything*) and she let out a huge breath and told me I wouldn't believe how much of a difference it made for her!

After teaching a recent meditation class one of the students who was a self diagnosed "worrywart" asked me what she could do to help remember a couple of the tips that she had taken away from the class. She explained that she needed help especially at work but that she didn't want her peers to know anything about her goals. My recommendation was a small rock, a tiny picture of something that she felt was serene or even just a little red sticker placed discretely in her office where she often looked - perhaps next to her phone. This would act as a reminder for her to stay in the moment and not get ahead of herself. She loved it and said she knew exactly what the item would be. Sometimes we just need to make it a little easier for ourselves. <u>A visual reminder</u>

Feel your heartbeat

"Each instant is a place we've never been." - Mark Strand

Perhaps those three words make you smile and want to excitedly close your eyes and take a moment for yourself. More likely, those three words make you want to roll your eyes at me and skip to the next phrase. Please give this a try.

You may have guessed that this is another technique used in mindfulness and in meditation. Being mindful is one of the most profound ways to change your life and meditation is the best way to get you there.

The really cool thing about this technique is that it's quick and it works.

After you've read this, I invite you to softly close your eyes and breath normally. Once you're in your own space and feel calm, gently place your attention at your heart center. Keep a curious mind and see if you can feel your heart beating in your chest. Can you connect with it? Don't rush or panic if you don't sense it right away - I promise it's still pumping. Just be kind to yourself and take your time. Some people are able to feel it immediately and then may elect to move to other parts of their body and feel the pulse in their neck, arms, hands, feet or anywhere else in your body. Others might possibly find it challenging to connect with their heartbeat and that's perfectly normal. If you feel that you're not able to get there, simply imagine your heart beating in your chest. Let go of any attachment to the outcome and just relax with your eyes closed in a comfortable position, imagining your heart doing its job and pumping healthy blood through your body.

This is something that you can do for a quick 30 second oasis or for an extended 20- or 30-minute meditation. It's an excellent way to slow down and give your nervous system a lovely little hug (or pat on the back, if you're not a hugger). Allow any thoughts or sensations to just pass by like clouds and gently bring your focus back to your heart center. Remember to smile when you're finished and take your time moving back in to action. **Feel your heartbeat**

Abandon the attachment

"When the power of love overcomes the love of power, the world will know peace." -

Jimi Hendrix

This is one of those life changes where you need to do some self-reflection in order to figure out what your attachment(s) is (are). The answers will be different for everyone however most people will have at least an inner whispering if not a full-fledged knowing of what it is.

This could be an unhealthy attachment to another person (no guilt or shame - most of us have been there), an attachment to food, material possessions, yourself (such as being overly attached to your looks, fitness level, education, etc.) or to any other myriad of things. Let's not kid ourselves, there's a lot of fun stuff on this planet to get attached to!

Many people have told me that they have an attachment to a substance. Abandoning the attachment won't necessarily be easy and you may need to find professionals to help you but the fantastic news is that recognizing the attachment is the first step to being able to leave it behind.

My son would immediately respond that his attachment is to his device. He loves games and even though we have always limited his playing time he knows that once it's running, he goes in to his own little world.

For me the attachment that I struggled with the most was the attachment to outcome; what I wanted to have occur. This may ring true for a lot of people and might look something like any of these:

- How will s/he respond to my text?
- What will my boss think of this work?
- How will I ever get this finished?
- Will my baby be healthy?
- Am I ever going to lose this weight?
- What if I'm not doing what I'm supposed to be doing in this life?

Anything that has to do with "what's going to happen" is an attachment to outcome. How can we not have it, if we care about the future?

Whether we care about it or not will not change things once they are out of our control. When the exam is over and handed in, there is no amount of worrying about your mark that will change the way that you responded to the answers. It will be what it will be. Accepting this as soon as you possibly can will make your life much more serene. Yes - serene.

I was employed as a server for many years in my late teens and early twenties. I would go to work every shift trying to "wish" the way it would turn out. If I was short of cash then I would put my energy in to hoping that the restaurant was "slammed" that day. Conversely, if I was tired from being out late the night before then I would hope with all hope that it would be "dead" at work.

One day while we were slammed for the lunch rush, I took a moment to look around at the situation. The servers were dashing around, some of them visibly sweating from the pace. The kitchen staff were either getting frustrated with the servers or having the servers be frustrated with them. My boss was barking orders like we were on the front line in battle. All of a sudden it hit me - no matter what I want the day to look like with regards to our customers, the flow would be what the flow would be. Aside from secretly locking the doors or posting a sign at the front that said, "Dead rats found in kitchen" there was absolutely nothing that I could do to change it.

All of a sudden, I relaxed. It was as though the tension in my entire body leaked out of my feet and deeply away from me. I felt light. I felt free.

I finished the rest of my shift with a calmness I wasn't used to and a stupid grin on my face. The customers came and they went - they got their good service, they got their good food and they paid their bills and they left. Then more customers came. The difference was that I wasn't panicking anymore or listening to that continuous voice saying, "I'm way too tired for this today." Instead I was just dealing with everything that came as it came.

Weeks later my boss commented on the fact that when everyone else was running around like a chicken with their head cut off, I was the only one that wasn't rushing - she wanted to know why. I told her that all of my "tables" had what they needed and me running was more likely to cause an issue than to help one. I explained to her similarly as I've shared with you that the rush will be what it will be and as long as I stayed calm and allowed it to happen that I would be in the best frame of mind to handle it. She looked at me pensively for a moment and then her shoulders visibly relaxed and she nodded at me before she walked away. I think she liked that. **Abandon the attachment**

Their own path

"We are here on earth to do good for others. What the others are here for, I don't know." - W. H. Auden

Like a lot of people, there are times that I wish I could be Ruler of the Universe. (Oh, the things we could do, "improvements" we could make...)

This phrase is about the art of letting go. We all have people in our lives that we want to behave in a different way. Whether it's a family member that we don't agree with, a colleague that we feel could use an attitude change or even a stranger that we would love to have act differently than how they do/are.

Here's the thing - we are all on our own path. We all have things to learn in this world and we can't force people to learn them any faster or slower than they are meant to. We can't love someone enough to make them lovable, we can't push someone enough to make them change who they are and we can't give enough to make them choose to overcome their addiction or pain. Although we can love, support and offer to guide them, their path belongs to them.

How much of your life do you want to spend fighting with someone to "make" them see your point? The fact is that there is free will and that everyone needs to make their own choices and decisions. We can do the best that we can to support them, but we can't change them - nor should we sacrifice ourselves or our mental health trying.

Years ago, a girlfriend and I were out for dinner. Gina told me that she was incredibly upset because she and her sister, who had always been close were no longer speaking to one another. When I asked her what happened she explained that her married sister Emily, mother to a son, was having an affair with a man at the office where she worked. Gina said that she kept fighting with Emily to end the relationship and that now her sister had cut Gina out of her life completely. She continued on in great detail about how terrible and bad Emily was and how her decisions were all wrong. When Gina had finished venting, I said, "That must be a very difficult situation for your sister. And with you gone, it must be very lonely for her as well."

You can imagine that Gina was not at all impressed with my comment and felt that clearly, I hadn't understood the point. I went on to explain, "Gina, you would consider yourself to be quite religious, right?" to which she responded affirmatively. I then asked her if she felt that there were a lot of good people in the world who were not religious and she again said, "Yes." I asked her to imagine how she would feel if Emily told her that religion was bad, that she was doing the wrong thing and that she didn't support her. Would she feel hurt, betrayed, torn, abandoned or possibly even resentful? Would she be tempted to stop communicating with her sister if every time they spoke, she was told how terrible she was for believing in God and religion? Gina slowly nodded. I explained that she can still love her sister without agreeing with her decisions - she can still support her sister without supporting her actions. I gently continued that it must be so conflicting for Emily to have to live two separate lives; how difficult it would be and what it represented may be happening in her life and marriage that nobody knew about or understood. While Gina was contemplating, I offered that she give her sister a call - tell her that she loved her, was sorry for judging her and was here for her if and when she needed. It doesn't mean that Gina was condoning her sister's actions or agreed with the decisions. What it means is that unconditional love is exactly that - loving someone without conditions, including respecting the decisions they make whether we agree with them or not. I know it's difficult, but sometimes love isn't easy.

The next time that Gina and I went out to dinner she thanked me for my advice. She told me that she and Emily had reconciled and were speaking again. They had agreed not to talk about the affair but were once again connected with the other aspects of their lives. Gina told me that I had reminded her about the lesson of love, which is a big part of her religion.

The affair did end months later but that's not the point of the story. The point is that Gina's sister felt strongly enough that the affair was something she needed to do and nothing was going to stop her. Being angry about Emily's decisions and actions would not have changed the outcome, other than the fact that the emotions would have eaten Gina up and the relationship might have been sacrificed.

Although sometimes it's really hard to accept, it's not up to us to judge, condemn, "fix" or change anyone. We can offer our suggestions and our love and then let them do what they feel they need to do. It's their life and not ours - we have enough of our own issues to deal with that we don't need to take on someone else's. Take a breath and release the need to be "right" and have someone see *your* point of view regarding *their* life. Not only is this a huge gift to them, it's an even larger gift to yourself. **Their own path**

Life gives souvenirs

"Mindfulness does not erase negative memories; it 'transcends' them, giving us back our deepest power which resides in our hearts." - Christopher Dines

How many times have we heard sayings like, "Life is what you make it" or, "Whatever you think - you're right"? Likely ad nauseam. The thing is that we've heard these sayings so many times because they really are accurate. How we look at things is how we perceive them to be, which means that it's what we believe to be true and that becomes what our life reflects. A cool thing about this is that we can easily change how we look at things simply by deciding to change how we look at things. It's honestly that basic.

How different would you be if you looked at your "war wounds" in life as "souvenirs" instead? Life is a journey and like so many journeys, sometimes we pick up souvenirs.

We have been given the gift of these tremendous bodies to live in while we're here on earth. Some of our bodies are bigger or smaller, taller or shorter, hairier or balder, healthier or less healthy, diseased or not diseased - you get what I mean. The fact is that they are what they are. Possibly we can change some of these things and possibly we can't.

I have many souvenirs on this body of mine and I love and appreciate them all. One of them runs the length of my left eyebrow. It was obtained from a particularly vicious attack one night as a teenager when my Stepfather went to punch me as I sat at the dinner table but had forgotten to remove the water glass in his hand before he landed the first strike. It's a scar that you can't easily see because the hair of my eyebrow covers it however when I go to get my eyebrows waxed, I often get asked, "Did you know that you have a scar under there?" (As a side note, that comment always makes me laugh because how could I possibly have a scar that large and not know it was there?)

Anyway, that souvenir is from something that happened to this body a long time ago. It didn't happen to *me* (the soul/person/being inside of this body), it happened to *this body*. The miraculous news is that I survived - and that I got a free souvenir to remind me. Do you see the difference between looking at myself as the victim of a terrible event versus the survivor? The outcome is the same - I'm still here. The difference is that I don't let it dictate my life or define who I am. If anything, this souvenir reminds me that I'm a fighter, I'm strong and I'm still standing. What a wonderful gift!

I have some beautiful souvenirs on my feet and they light up for me every summer after they get some sun on them. They were obtained when I was very young and fell in to a bathtub of scalding hot water that had me enjoy an extended stay at a local hospital.

When I was on a house boating trip with a group of women in my twenties, I went up to the top platform to bask in the sun for a bit. The lone woman next to me had souvenirs on her feet that were just like mine. I told her that I loved her scars - they made her feet so unique! She was embarrassed and quickly went to cover them until I showed her my feet and said that hers showed better because of her beautiful dark skin. When I put my feet next to hers to compare, she opened up and told me her story. When she was a child she was placed in a tub where the nanny had put in the hot water without also adding the cold water. She told me about the skin graphs from other parts of her body to cover her feet and showed me those souvenirs as well. I told her that her story was incredible and truthfully said that I thought she was amazing - and that she had marks that nobody else in the world had. (I also said that her story was way more exciting than mine..although I did get to ride around in a wheelchair and ate orange popsicles at the hospital until they ran out.) She switched from being ashamed of her souvenirs to being proud - and rightly so! What a story and what a survivor. I still think of her and smile. What a strong and special lady.

Some of our souvenirs are private, yet some of them are available for the world to see. I've met many people who have birthmarks or other souvenirs that they themselves consider to be "unsightly." Sometimes strangers have even made comments about these marks. It may not be something to cherish at the time, but if you look at in the grander scheme, perhaps these souvenirs have given other people that make mean comments the gift of hindsight and learning. Perhaps it taught them a lesson in humanity that prompted them to change. We have that person with the souvenir to thank for that.

Everyone has souvenirs. As an example, so many of us have badges of honor through the use of stretch marks from extreme weight loss or pregnancy. Instead of being ashamed of them, be proud of them - you're different, you're unique and you have a story. The story doesn't have to define you, although it can be a part of you. The part that survived. **Life gives souvenirs**

Forget the math

"If there were none who were discontented with what they have, the world would never reach for anything better." - Florence Nightingale

Most of us have done it - the math works but it doesn't feel right and we venture forward anyway. Why do we discount how we feel because something looks good on paper, to others or even to ourselves?

Allison was a close friend of mine who had met her partner Jeff at the marketing firm where they both worked. Jeff was a really good guy. He was attractive, had a great reputation, a job that paid well and he adored Allison. They were "the" couple at work, depicting a perfect match at a perfect company. The issue was that Allison couldn't shake the fact that she just wasn't feeling anything for Jeff. He said and did everything right but she explained to me that being with him just felt wrong for her. She did the math again and again and again - it worked...this should have been the perfect relationship. Allison didn't want to disappoint their families or all of their peers who found their relationship inspiring. Against her better judgement Allison continued to stay in the relationship hoping that eventually she would feel some sort of deep connection or even attraction for Jeff. It never happened and ultimately highlighted the fact that sometimes you need to just forget the math.

You can consider this in any multitude of ways; yes or no, right or wrong, good or bad, green or red or one of my personal favorites - yum or yuck. Although Allison tried to convince herself otherwise, the relationship had been a "yuck" experience. While nobody could understand it, she needed to terminate the relationship with Jeff - and what a huge relief for her when she did.

The concept of "forget the math" is meant to be easy and simple - the question is, "When nobody is looking, how do I *really* feel about this?" You may be shocked with what you find.

Give it a try; close your eyes and choose anything about your life that comes to mind. This may include relationships with partners, family members, friends, co-workers or others and it may include your beliefs in life or how you truly feel about your job. The list might also involve your money situation, volunteer activities, parenting, schooling or any other host of alternatives.

Now ask yourself, "Deep down, if nobody will ever know, how do I truly feel about it?" If it's easier, just ask yourself if it's a yum or a yuck. The answer may surprise you.

If something comes up for you as merely your "cross to bear", that may be worth further investigation. Life paths are not meant to feel like an obligation. The one with the choice and power to change your life is you.

Remember that what you find doesn't mean that everything has to happen right now. If you dislike your neighbors it doesn't mean that you should put your house on the market tonight and if you realize that sometimes being a parent can really suck it doesn't mean that it's time to phone social services. What it means is that now you know how you truly, authentically feel. That's the first step to any change. You may find that it's an easy move to make or that it's best to make your next move through the use of professional help.

Good luck. Forget the math

Take the step

"What saves a man is to take a step. Then another step. It is always the same step, but you have to take it." - Antoine de Saint-Exupery

So many of us at certain times in our lives yearn to be more than who and what we are. Yearning to be better is a natural and innate desire for growth that helps humanity to progress. This desire is a crucial step to ensuring the sustainment of our species and is something to be welcomed with open arms and celebrated.

At one point in our lives, we need to make the conscious choice to be what we would like to become. We can't simply wish for ourselves to end up being the person we always dreamed we would be without taking that first step to being said person.

If you wish to be remembered as someone who was incredibly loving then you need to make the effort to release fear and give love. If your desire is to be known for being a philanthropist then you must make the first move towards giving to others. If being identified as a pioneer is important to you then the choice is to take a deep breath and break new ground. It's that simple. And it can be terrifyingly scary.

Depending on how deep the canyon is between who we are and who we wish to be, that first step may be as scary as certain death. In a way, perhaps there would be a death; that of who we are in order to give birth to who we want to become. How badly do you want it?

Just as the caterpillar must release his attachment to who he is in order to become the butterfly he was born to be, sometimes we too must release our attachment to our ego or our old self in order to fulfil the destiny we wish to have.

There is empowerment in growth, especially when it is growth that we choose versus growth that is thrust upon us when we are not ready or willing.

Become who you are ready, willing and wanting to become. *Take the step*

I love myself

"No amount of self-improvement can make up for any lack of self-acceptance." -

Robert Holden

A few years ago, one of my clients, Elise told me about an appointment that she had recently had with one of her long-time clients. They sat down at Elise's desk and the gentleman slid a regular letter sized piece of paper across the desk to her without saying a word. She picked it up and noticed that repeated over and over again was the hand-written sentence, "I love myself." It filled up the entire 8 1/2 x 11-inch page. She looked at him quizzically and asked, "What's this?", to which he replied, "Elise, I need to write that every single day so that I have the strength to get out of bed in the morning."

This gentleman was suffering tremendously from depression and needed personal affirmation on a daily basis that he loved himself. He needed to know that he was important and that he was worthy. Every single person on the planet needs that.

If we all know that it's imperative to love ourselves tremendously and we all know that we will be better human beings if we do, then don't wait another day to start up with the life-saving affirmation that Elise's client had.

Look at the contribution his painful journey would make to the world if even one person learned the lesson he has to teach.

Every single day, remind you that you love yourself.

For some of us this is extremely easy, in fact some people may be confused as to why this phrase even made its way to this book.

For others the thought of loving oneself is almost unfathomable. I have met countless people who have shared with me that they don't even like themselves, let alone love themselves. That's a long-haul because you're stuck with you for the rest of your life . The relationship you'll have with yourself is longer and more intimate than any other relationship you'll ever have with anyone else. So why not work to make it a good one?

It's remarkable that as children we're taught how to talk lovingly to others but we're not taught how to speak lovingly to ourselves. We're taught how to engage respectfully with others, but were we ever taught specifically how to be respectful with us? Hopefully we're taught to connect with those who care for us but we likely weren't taught how to connect with ours most inner being. How is it possible that these vital lessons have been missed in western culture?

It is not too late to learn now and start integrating them right this very moment.

You are magnificent. You are special and you are unique. Even with six billion other people on the planet, there is nobody else just like you; you are an original. Your imperfections are what help to make you so incredibly perfect. Nobody on this planet can be you better than you can.

Make it a habit to think of at least three things every single day that you love about yourself. I promise it will help with so many aspects of your life, from confidence and more peace to better relationships with others. It's miraculous what happens when you love yourself to bits.

If this notion is overwhelming and you're thinking to yourself, "There is literally not one thing that I love about me" then you are not alone. (Just as if you are thinking that this is silly because you already love yourself more than anyone in the world, you too are not alone.) You are never alone.

If you're having trouble coming up with a grocery list of things that you love about you then you may find that sitting in front of the mirror helps. The only rule for mirror work is that there is absolutely, unequivocally NO negative self-talk allowed - EVER.

Is there something that you do with work or your personal time that makes a positive difference to others or the planet? Are you an expert at turning off the water tap while you brush your teeth? Does it make you feel good when you recycle? Do you donate to a charity that does important work? What's awesome about your hair? Your freckles and moles are so cool! Are you blessed with some special birthmarks that make you distinctive? Maybe you've never noticed before how adorable your earlobes are or how perfectly your mouth fits on your face. You've got a great laugh - what makes it great? That it's infectious, humble, silly, that it originates in the gut or comes out like a snort? You have an adorable pinky toe. Your height is perfect for you. Your scars tell a story of survival and resilience. Nobody has eyes like you and when you look deep in to them, they see in to your soul. Help them learn that what they see every single time they look at you is love. I love myself

Life after trauma

"As I walked out the door toward the gate that would lead to my freedom, I knew if I didn't leave my bitterness and hatred behind, I'd still be in prison." - Nelson Mandela

Regrettably this planet has a lot of opportunity for beings to experience trauma. The statistics are staggering and deeply sad regarding people that will suffer from trauma in their lifetimes. Trauma could include physical, sexual or emotional abuse, the death of a loved one, witnessing a traumatic event, a severe illness, being bullied, surviving a natural disaster or accident or a huge list of other emotionally or physically harmful events. Unfortunately, a lot of humans will suffer trauma in their lifetimes and many people will experience more than their "fair share."

Life after trauma is very different than living after trauma. Anyone who has suffered from trauma and has survived is still living but they may not feel as though they are alive. Trauma victims who have not fully healed often feel as though they are going through the motions of living but not *actually* living. Life is possible with the right help, support and knowledge. Mindfulness is a necessity to living fully after trauma.

When I was 13, I participated in a roll-a-thon at a local roller-skating rink to raise money for our band class. It was an event that had us skating overnight and the music played loudly to keep us all energized and moving. There were a number of us girls all holding hands as we skated that evening and the friend whose hand I was holding let go and skated in to area with the benches. I figured that she was going in to take a break and have a drink so I kept skating. What actually occurred was that she was having a severe asthma attack and went to get her inhaler. She died that day.

As I sat in the pew at her funeral, I was acutely aware of how profoundly shattering the entire event was; her death, the heart-wrenching eulogy from her sister and the indescribable pain and suffering of her mother as she sobbed uncontrollably.

I missed my friend and felt overwhelming guilt for not asking her why she was leaving or for not following her to the benches. I fully and wholly blamed myself for her death.

By that point in my life, my trauma defense mechanism had already kicked in to high gear and I was incapable of emotion. I sat in that pew wanting desperately to cry but not one tear would come. I felt like a monster; as though I was inhuman and it was just one more thing to add to list of what a shameful person I was.

Trauma affects different people in different ways, all of them our brain's way of trying desperately to protect us and make sense of things that have happened. As I look back, I am tremendously grateful for the protection strategies that my brain implemented for me as a child so that I could cope with, and one day overcome my childhood.

Without fully healing, there is no time-limit on trauma. It's not as though you can look at your calendar and say that an amount of time has passed so now you should be feeling better. If you're not healed, you're not healed and although time may possibly help to reduce the effects it is highly unlikely that it will fully erase them.

Symptoms of trauma can include anything from having a hairpin trigger with anger (even for the seemingly most innocuous things), depression or anxiety, self-punishing behaviors including addiction, self-cutting and eating disorders, not being able to concentrate or remember or the inability to feel any type of emotion and everything in between. I have had more than one person tell me in confidence about the guilt they felt by having absolutely no emotion when their baby was born. Although it could indicate a lot of things it could also be a sign of repressed trauma - *not* a sign that they're a terrible person, which is how they were feeling.

Trauma is like a wound to the soul and like any wound it must be acknowledged and appropriately cared for.

There are too many people reading this book who have suffered trauma. There are too many people reading this book that did not get the help and support they needed to overcome the trauma that they suffered. Perhaps they bravely thought they could "tough it out" or they were in denial. Perhaps it was a lack of resources or understanding or any other host of reasons. It doesn't matter; what matters is getting help now so that you can fully enjoy the richness of life that everyone on this planet deserves.

Imagine a life where you feel fully; you are engaged and feel happiness, joy, gratitude and love. With it, also feeling sadness or sorrow, anger and hurt but knowing that you have the strength to move through it. Imagine truly intimate relationships where you can be honest and can trust your loved ones. A life where you don't feel guilt or shame for something that wasn't your fault or where you can find forgiveness if it was. It's there just around the corner - find the proper support and take your first step to getting there. I can tell you that the path isn't easy but it sure is worth it. Perhaps the divine purpose of trauma is to give us the opportunity to heal. We do that by lifting the veil of our mind in order to discover and reconnect with our soul.

One of my most favorite books about trauma and healing past it is called *The body keeps the score* by Bessel Van Der Kolk. There are so many amazing books out there about trauma and reading them along with getting professional support is a great start to healing your wounds.

Your magnificent life after trauma is possible; it's just waiting to be discovered. Life after trauma

Forgiveness through compassion

"The hardest forms of compassion are for people you don't love." - Paul Gilbert

One of the most impactful gifts that we can give in this life is forgiveness. This may sound trite but please love yourself enough to give this concept a moment of consideration. The gift of forgiveness is for the person you are forgiving, for the world as a whole and most importantly for yourself. There is a saying that hating someone is like drinking poison and wanting the other person to die. The person we harm most is ourself. I have personally seen this time and again with people who are so full of anger and hurt. The hardest part is witnessing in them that holding on to all of that hate and negativity darkens the person holding it so much more than it will ever darken the person it is intended for. You deserve so much better than that.

Although there are many events in my childhood that forever shaped me and my personality, one that sticks out the most occurred when I was five years old sitting in the back seat of our Cutlass Supreme Oldsmobile arguing loudly with my siblings. My mother decided that she'd had enough and stopped the vehicle. She asked me if I knew where we were and when I replied that I didn't (as we were many miles away from home) she flatly told me to get out of the car. I did as I was told and she drove away at full speed.

Initially I was defiant, refusing to have her see me upset by her latest terror tactic however the severity of the situation quickly hit me and I realized that I would be alone and lost with nowhere to go and nobody to help me. I was literally terrified. My mother had yet again abandoned me and this was the most public display of it yet. Fear gripped my body and I started running down the street, crying and screaming for my life, flailing my arms in an attempt to have her see me in the rear-view mirror. My ego was cast aside for the greater aspiration of once again being "safe."

The car stopped before turning the corner and utterly humiliated, I slid in to the back seat at which point one of my siblings whispered, "I bet you feel stupid." I did.

At five years old I silently vowed to never, ever let anyone see me flinch again. It was a promise I kept for the remainder of my childhood.

At age 12 I had the opportunity to relive that experience. One Friday afternoon my mother was angry with me because I had stopped to taste a sample at the grocery store while she was shopping. After loading the groceries in to the trunk she got in to the car and refused to unlock the door for me. She reached over and rolled down the window to ask me if I knew how to get home. I told her that I did not at which point she promptly drove away. Inside of me raged a deep anger and resentment towards her; a powerful and fiery inferno of hate and I refused to allow her to see that once again she had left me helpless, scared and alone. This time she would not see me flinch.

I started walking through the parking lot, drinking the poison of hate. I wished that I would somehow end up dead that day; murdered and left in a ditch. After some thought I figured that would work out too well for her because then she could play the innocent and grieving mother, some lie about why I had been out there alone. I refused to give her the satisfaction of having me dead.

Sometimes the universe works in beautifully mysterious ways and not too long after a neighbor drove past me and parked close to where I was walking. Although she was shocked to hear that I had been abandoned, fortunately for my mother this neighbor was no stranger to abuse so there would never be mention of this incident to anyone.

The part about that story that makes me shake my head the most is that when I did make it home hours later, after my neighbor had finished her shopping and errands, my mother was not there. She and my stepfather had gone away for the weekend and did not return until Sunday night. Although my mother didn't call or communicate home at all for those 48 hours, she did assure me when prompted that she "felt bad" and "wondered a couple of times" if I had made it home while she was gone. In fact, she made sure that I was aware of the fact that it had put a bit of a damper on her weekend away.

This is one of many stories of my youth that I could choose from, however it makes the point I am trying to make. The person I was hurting the most when I was busy hating her was myself. The poison ran through my veins and made me dark inside.

Afterwards, in trying to make sense of the incident I came to the realization that no mother in her right mind would ever do to us the things that my mother did. My mother was not in her right mind. That epiphany was a game-changer for me. It was true; what good mother on this planet would ever do any of what she had done? Nobody ever has children with the intention of being a narcissistic parent. I actually accepted the fact that she did the very best that she could. Her best is different than what my best as a parent looks like, however why would she have done anything other than her best? She had her own hurts and traumas from her childhood that helped to shape her in to what she had become and I felt sorry for the little girl in her past that was also abandoned and alone, unable to heal.

This realization changed my life. The anger and hate I felt for her slipped away and at 12 years old I became thankful for the lessons that she was teaching me. I was learning that I could fend for myself. She was teaching me about the parent I wanted to be when I grew up. Although I didn't know it at the time, the experience allowed me once again to be a survivor versus a victim. It was as though the weight of the world had been taken from my shoulders and I could feel the sunshine on my skin again. The sweetness of forgiveness through compassion filled my entire body.

Now, in no way do I condone her behavior. In no way is she absolved for being a neglectful parent. The only thing that changed was my perception, my anger. What changed is that I forgave her even though she didn't ask for my forgiveness. I gave my life back to myself by feeling compassion towards a mother that would hurt her children as deeply as she had hurt all of us.

I have love for my mother as I have love for all beings. However, after many more years of terror tactics and abuse I finally made the decision to not have her in my life any longer. Although I wish her all the very best and have no feelings of ill will toward her, she has not been in my life for a long time, nor is she in the lives of my children. I am at peace with the fact that nobody could protect us kids from her back then but feel that it is my duty as a parent and a loving adult to protect my children from her now.

Many people may disagree with my decision. I fully respect that. I had to do what was right for me and you will do what is right for you. What you decide to do is not for anyone to judge except for you and your heart.

Just because you forgive someone doesn't mean that all is well between you and you're going to be besties now. Forgiveness is not a synonym for absolution. Sometimes that can be the case and sometimes it can't, and possibly shouldn't be.

To truly put yourself in the other person's shoes, to feel compassion for them and then to fully forgive them is one of the most beautiful things that we can do during our time on this earth.

Give yourself the gift of forgiving someone, even if you never tell them.

You deserve it. <u>Forgiveness through compassion</u>

For future you

"My friend, let's not think of tomorrow, but let's enjoy this fleeting moment of life." -

Omar Khayyam

A few years ago, I had been up in Banff, Alberta for a three-day work conference and I was being driven back to the Calgary airport by a close friend and colleague. Kevin was sharing with me a number of items that were on his mind, some things that were heavily weighing on him and causing him some grief. At one point he stopped almost mid-sentence and after a pause and a long, deep breath he announced that these things were actually for Future Kevin to worry about. "Future Kevin?", I asked curiously. He explained to me that he had a habit of getting ahead of himself and that when things were overwhelming or he became anxious he would ask himself if it was an immediate problem. If it was, then it needed to be solved and dealt with, however if it wasn't, then it was for Future Kevin to deal with and didn't need to bother Current Kevin. He quickly let the concern drop and our conversation changed to the breathtaking Canadian scenery.

How would your life change if you incorporated this practice? Imagine an automatic cease-fire so that when anxious thoughts about the future invaded your peace of mind you simply put them back in their place by reminding yourself that it's for Future You to deal with.

What would you do with all of your spare time?! How amazing that you could make the conscious decision to focus your energy on Current You; the one that matters - the one that can actually change and shape your life.

It's still vitally important for us to do all of the day-to-day things that need to be done, we still have to plan and prepare but there is a big difference between gathering, collecting and storing our nuts to prepare for winter versus sitting idly panicking about what we're going to eat when the snow starts to fall.

Your real life is right here, right now - in this very moment. You can't change the past and worrying about the future isn't going to get you what you want. (Unless what you want is unnecessary and harmful hormones and chemicals coursing their way through your entire body - then hey, fill your boots.) Face and deal with all of the things that Current You needs to deal with and leave the stuff that you can't change by worrying about it to Future You. You may find as I did after taking Kevin's advice that most of the time Future You didn't even need to worry about those things in the first place because the fears and concerns never came to fruition. Variations to this could include, "Stay right here" or, "In this moment." *For future you*

Honor your body

"He who lives in harmony with himself lives in harmony with the universe." - Marcus Aurelius

One of the fundamental pieces of mindfulness is body awareness. In order to have an intimate relationship and knowing of ourselves, the journey often begins with an intimate relationship and knowing of our bodies. We've all got a miraculous body that allows us to be part of this earthly experience. We get one life with this body and it's the only one that we'll have while we're here. Why on earth do we insist on taking it for granted or trying to push it to the limits? When you've got a car and it's not treated well there is an option for you to get another one upon its breakdown. There is no option for a new body in this life - it's the only one we'll get.

From wetting our pants because we were having too much fun to bother going to the washroom (likely when we were younger but there's no judgement) to all-night binges and everything in between, most of us have had the opportunity to *not* give our bodies the respect and attention that they deserve. That's OK and you're here now (along with your body) however it's a perfect time to count our winnings and walk away from the abuse-your-body table.

Our bodies are incredibly intricate and profoundly organized. Like any object, the better you treat it the better it tends to respond. (Of course, there are always exceptions and we make the most of what we have.)

By no means am I recommending that you start training for your first or next Triathlon tomorrow. Why not start a little simpler with mindful guidelines such as:

- I only eat when I am hungry
- I stop eating when I am full
- I eat food that nourishes my body
- I empty my bladder when I feel the urge to urinate (instead of "holding it" for hours on end)
- I sleep when I am tired
- I do not drink alcohol or I stop drinking alcohol when I have reached my limit (and I decide my limit *before* I start drinking)
- I limit my caffeine intake
- I honor my emotions by identifying and acknowledging them (and get support if I need help in dealing with them)
- I clean myself when I am dirty
- I smile because it feels good
- I make healthy decisions
- When I am stressed and rushing, I slow down instead of "pushing through" and speeding up
- I take regular deep breaths and fully enjoy the exhale, as it calms me down (by engaging my parasympathetic nervous system)
- I exercise at a level that is good for my body and stop when my body tells me that it is time to stop
- I nourish my mind with positive thoughts and emotions
- I take the steps necessary to ensure that my digestion and elimination processes work the best that they can
- I hydrate with water when I am thirsty
- I find as many reasons as I can to have a deep, nourishing laugh

- I regularly engage with people in my life that help me to feel safe, secure and loved

You get the gist. There's lots more to choose from and the best person to decide what your body truly needs in each moment is you. These aren't things that make you want to throw your hands in the air screaming, "I can't do it!" They're things that you read that make sense to you - so simple, so easy, so smart yet regrettably so easy to ignore and push away.

Honor your body as though it is your most special and unique treasure. Honor your body as nobody else can. Honor it because you love yourself. Honor it as though it's the only one you'll ever have. Honor your body

Take baby steps

"Great things are not done by impulse, but by a series of small things brought together." - Vincent van Gogh

Many of my closest and dearest friends have suffered from depression and/or anxiety. A best friend and soul mate during my life in Toronto lives with Obsessive Compulsive Disorder (OCD) and Bipolar Disorder. I'm personally thankful to have lived through my dangerous struggle with anorexia in my early teens. Countless employees and clients over the years have confided in me that they or their family member(s) were currently battling, or had in the past wrestled with mental illness. If you aren't personally suffering from a mental illness, statistics show that you know someone who is. This is truly an epidemic that will only be cured through knowledge, compassion and community.

For what it's worth, my personal finding is that sometimes people who have lived through a mental illness can come across as more real and authentic than others who are "perfect" because they've had a deep, profound and true look at themselves and their history. They had or were about to have their own version of the "dark night of the soul" and successfully come out of it the other end.

Over dinner one night almost 20 years ago my girlfriend and colleague Sharon shared her story with me. She told me in detail about being abandoned as a child and never feeling safe or loved. She talked about an episode at work that pushed her over the edge and caused her to need hospitalization in a mental facility for months and about the devastating and immobilizing crying spells that still gripped her and would last for days at a time. Her story was raw and moving.

Sharon's narrative was unique in the details yet alarmingly similar to so many other stories that I have heard. Most of us have had the opportunity to survive through very dark times. We humans have an innate desire to survive and our resilience is remarkable.

Whether it's a new or lingering mental illness or a horrific life-changing event such as the death of a loved one, conclusion of an important relationship, health issue or other traumatic event, the simple advice that I have given and used for times of despair is, "Take baby steps." When it feels like you can't make it for another minute let alone another day, take baby steps. It is a moment-by-moment message of love and encouragement to yourself. You don't have to run before you can walk and you don't have to take full steps before you take baby steps. After trauma occurs its time to learn again - find a new you and a new normal. Be kind and encouraging to yourself. When things are so bleak and hard (and I wish with everything in me that nobody ever has a moment like this again) then just take baby steps.

Upon waking up in the morning if it feels like it's all too hard and overwhelming, take baby steps.

And after you go pee and you don't know how you can make it through another second then just take baby steps.

And then put your clothes on one article at a time and take baby steps.

You can do it.

It's heartbreaking not that I feel compelled to share this advice with you, but that it's so needed because sometimes life is really, really hard. I get it. So many people do. I promise you that you're not alone - the mind just tricks us in to thinking that we are. That's part of the illusion of mental illness.

Help is out there. Keep looking if you haven't found what's right for you and don't give up.

The incredible thing about getting through this is that one day you will. And one day all of a sudden you realize that you're not having to say it to yourself very often. And one day you realize that you don't have to say it at all, because you did take baby steps and you did get through this.

Keep going. Find something to be grateful for and when you're done that, find something else to be grateful for. Celebrate every single thing that you do - celebrate that you got the kids off to school with healthy lunches. Congratulate yourself for everything from walking to the mailbox to asking for that raise or whatever else your goal was. You got this. You can do it, whatever it is. One step at a time. Take baby steps

Choose to thrive

"There are two ways of meeting difficulties. You alter the difficulties or you alter yourself to meet them." - Phyllis Bottome

We all have stories of people who have wronged us in this lifetime, some of them more terribly than others. It's normal to have emotions of pent up anger and even hatred, and to have thoughts of revenge.

However, just because it's normal doesn't mean it's right and it certainly doesn't mean that it's good for you.

So, you'd like to say something gruesomely hurtful and venomous to them in order to help them to understand how terrible they are? Will that get you what you want; an apology, your sense of well-being, your money, your time, your heart? Likely not.

You have fantasies of hurting them in some way? The price to pay for that among many other things including karma is getting caught.

And even if you knew that you could do something insidious without ever getting caught - would you be OK to live with that for the rest of your life once the emotions had released and all you were left with was the knowledge of your hateful actions?

Forsake the revenge. Forsake the hurtful, hateful words. Consciously choose instead to thrive.

Live your life in a way that makes you truly, wholly happy. In a way that makes you feel like a good, worthy person. Let them have the karma of their actions and move on. Smile. Laugh. Love. Exercise. Enjoy food, enjoy reading, enjoy exploring, enjoy you - enjoy life!

(Besides, over the long run that will drive them absolutely *nuts* anyway.)

Allow that other person to hold on to the energy of everything they've done to you or said about you. And allow yourself the freedom, the gift, the love of enjoying your life and everything in it. Make the choice to take in and fully absorb everything good that this life has to offer. Choose to thrive

Everyone leaves ripples

"Everybody talks about wanting to change things and help and fix - but ultimately all you can do is fix yourself. Because if you can fix yourself, it has a ripple effect." - Rob Reiner

As one of many Scout Leaders in our group we wanted to teach the young Cub Scouts and Beavers about the way that we are all connected and can easily affect one-another. The age appropriate story that was shared was of a child who was stuck with a "red anger dart" through something unpleasant that had happened that day and that they then stuck another child who stuck another child and soon everyone in the group was stuck with red anger darts. The solution was to realize that you had been stuck with a red anger dart, to remove it, heal the wound and to choose to loft rainbow balls to others instead. It's the young version of the adult concept about how all of our actions create ripples that affect others.

Just as a small pebble dropped into a pond can create ripples that continue for great lengths, so too can the small actions of each of us have far reaches.

Working in banking for so many years and as a server for many years before that, I had the opportunity to see a lot of people mistreat others. Sometimes clients would fight with each other over their spot in the line up, sometimes they would yell at staff if they didn't get what they wanted and at times they would complain loudly if things didn't progress as quickly as they would have liked. Those one or two people would send out red angry darts to everyone that witnessed their behavior. The likelihood is that many of the people who had been struck with those red angry darts passed them on to others, even if just through the re-telling of the event. Those ripples went a long distance and who knows where and when they stopped. What's concerning is that those events and much worse happen all day, every day all over the world.

On the other hand, sometimes clients would be so patient with one another. Sometimes they would let others in line ahead of them for various reasons and sometimes they would reassure staff that the lineup, "wasn't too long at all." At times people would even go above and beyond to make others feel good or make their day. Their ripples lifted the hearts of everyone that got to enjoy participating in, or watching the experience.

Everywhere we go and with everything we do, we leave ripples. Gestures of kindness leave ripples that spread just as gestures of unkindness do. I like to believe that kindness ripples have a reach that goes even farther than ripples that don't include kindness. We can change the ripples of others through changing our own ripple and maybe something miraculous could occur. We'll likely never know the true effect of our ripples, but doesn't it feel good to be a part of it just the same?

Do something kind for someone you know or someone you don't know. Do it whether they'll ever know it or not. Choose patience over frustration and love over hate. Believe that it makes a difference, simply because it does. Everyone leaves ripples

Touch is necessary

"If you can learn from hard knocks, you can also learn from soft touches." - Carolyn Kenmore

I have patiently waited over 20 years for the book *Touch is necessary* to be written by a brilliant and inspired doctor or researcher. I imagined that the book would encompass page after page of research, data and personal stories of struggle and triumph. It would also include life-changing exercises for people who struggle with touch to regain their true sense of self and indulge in what is and should be one of life's most rightful joys.

Regrettably, that book has yet to be written.

In the meantime, let me passionately implore you to take pleasure in the safe and comforting touch of those that you love and trust, including yourself.

The research and information regarding touch starts as early as babies who are proven to be healthier both emotionally and physically when lovingly touched. This is one of the many reasons that health care practitioners often encourage mother and baby to engage in skin-to-skin contact as soon as possible after birth. Although we may outgrow the need for a lot of things in this life, touch is not one of them.

As mammals we instinctively give and receive comfort through touching and close contact with others when we are safe and respected. From holding, rocking and gently swaying a new-born baby as s/he cries to hugging a friend in need, and the intimate act of love-making with a gentle partner. Our nervous systems respond positively to healthy touch.

There are many times when touching can be very difficult for people. This is especially true if we have experienced a trauma, including being touched in a disrespectful, hurtful or abusive manner. Sadly, this is the case too much of the time.

Humans need to be touched. It is one of the most impactful ways that we exhibit love, tenderness and closeness with another. It is a way for us to provide and receive comfort and it calms our nervous systems when in need.

A good friend of mine was in a long-term relationship where his partner did not like being touched. Although he was incredibly supportive of his partner's needs, he also tried for many years to express what he was looking for and what he needed in the relationship. He was consistently the recipient of the response, "You know that's not me" along with a refusal to even consider trying to change. After many years together, he made the painful and difficult decision to exit the relationship and has now found a partner that fulfills his deep desire for connection and intimacy.

Although many of us may feel just as his ex-partner did that touching "isn't them", it may be an opportunity for us to gently and mindfully question why. Perhaps there has been a point (or perhaps many) where we were taught that touching can be painful, abusive, dangerous and violating. This should not be the case and touching in its purest form should be for us to heal and comfort one another and ourselves versus ever to hurt.

It is possible to enjoy touch. Proper touching can bring you to a closeness and overwhelming feeling of love and being loved that often words are not able to provide. It needs to be the right touching with the right person and for those of us who need it, we owe it to ourselves to find help in learning how to get there. Touch is necessary

Everyone has scars

"Our key to abundance in every area of life is this: We experience God's peace and harmony to the extent that we love, forgive, and focus on the good in others and in ourselves." - Marianne Williamson

It's so incredibly easy for us to look at one another and be judgemental. Whether we like to admit it or not, it's something that most of us do all day long, often without even being aware that we're doing it. We learned it from childhood and it's stuck with us like a silent disease.

We judge that person in front of us for driving like an idiot, the person in the line-up is rude, the person across the street should learn how to dress better, a family member should lose some weight or live their life differently. Constantly, consistently - we are judging.

Throughout my years of working with clients, I have learned many things and one of them is that every single person I have ever met has scars. Often times it is difficult (if not impossible) to know their scars just from looking at them, but trust me - they're there.

Shirley was an older client that I found very difficult to make a connection with. I felt that she was consistently standoffish and resistant of me. One day when I walked in to Shirley's office to meet with her, I was overwhelmed by a feeling of deep sadness in the room. Before we began, I looked at her and said, "I know we're not close and I completely want to respect your privacy but, are you OK today?" She looked intently in to my eyes for a moment and then let out a big sigh. Slowly she told me that her father's health was failing and she was torn as to how she felt about it. He had been abusive when she was a child and she had ended up in a harmful marriage because of it. She was at a point where she was facing everything that she had struggled with in her entire life, including her divorce and the aftermath of her children for having left their dad.

After reflecting, I offered that she must be proud of herself for being so strong. That she was an amazing survivor and had broken free from the chains that had bound her in the past. She was fortunate to be at a point in her life where she could choose, as a healthier person, how she would respond to the situation of her father's declining health.

This "crotchety old lady" was incredible and I had almost overlooked her because of my own egocentric need to judge.

Another one of my clients seemingly had it all; everything that people admired. Lisa was beautiful and slim, successful and popular. What people didn't see was that because she had always been judged on her looks, she was incredibly insecure and anxious. She had a deep-rooted fear that when the looks faded her self-worth would disappear as well. Although Lisa had always been judged in a "favorable" light, it was suffocating her to be judged as beautiful. She was scared every single day and believed that if she didn't have her looks, she didn't have anything. Judgement hurts everyone.

Maybe that "idiot" driving in front of us is rushing their sick dog (and best friend) to the vet. Perhaps that rude person in front of us in the line-up didn't even notice that they cut ahead because they're having a panic-attack due to their wallet being stolen the night before. Possibly the person across the street is dressed as well as they can be after taking what was left of their Social Assistance cheque to buy the only outfit they could afford for their job interview this afternoon. Maybe that person in your family didn't want to worry you that their weight gain is due to a health issue.

Who are we to judge when we don't know anything about anyone, including the people that we sometimes think we know? It's not hard to give someone the benefit of the doubt when you try. It's a gift to everyone involved, including you. Something that you can bank on is that there has been something in their past that hurt them. Everything that they have been through has made them what you see today; "right" or "wrong." We're all just trying to do the best we can with the scars that we have. And the world will be a better place when we practice compassion over judgement. **Everyone has scars**

Mid life awakening

"Wherever we are, it is but a stage on the way to somewhere else, and whatever we do, however well we do it, it is only a preparation to do something else that shall be different." - Robert Louis Stevenson

Around the age of 10, I was with my Mother at her friend's house pretending to be absorbed in what I was busy playing with so that I could listen intently to their adult conversation. Celia was crying, talking about her husband; he spent way too much money on a new fancy sports car and then left her. My Mother held Celia's hand and reassuringly told her that this was just his mid-life crisis and that he would soon be back.

I still remember my thoughts from that afternoon. "Mid-life crisis?", I asked myself. It didn't sound like a crisis to me; he bought a car that made him feel good and left a relationship that clearly for whatever reason wasn't working for him. It struck me that this seemed more like a "mid-life awakening".

Fast forward 15 years and it's time for me to have an awakening of my own. Because I was only in my mid twenties at the time (and my plan is to live to be a healthy Centurion) I thought of it as my Mid mid-life awakening. I had been in a committed relationship for almost 10 years - married and mortgaged, when I deeply and painstakingly questioned if my life was heading in the direction that I wanted. My career was on track however I no longer felt for my partner as I did when I was that teenager who had agreed to marry him. He was my very best friend and I loved him tremendously however we had both changed so much that it became painfully apparent if we were to have a first date at that point in our lives, there would not be a second.

I fought with the shame I felt for no longer being in love with him. I tried and tried and tried but couldn't force myself to feel how I desperately wanted to. I fought with the pain of after so many therapy sessions having to admit that no amount of counselling was going to get me to where I wanted and needed to feel about my husband. Oh, the humiliation and embarrassment over no longer being the perfect high-school sweethearts couple. It was a dark, difficult and lonely time.

It was also a time for change. I was tired of being so responsible and careful, tired of being routine and boring and absolutely exhausted from forcing myself to be someone that I no longer was. Within the span of a couple of years I bought a motorcycle, pierced my belly button, finally pulled the trigger on my "failed" marriage, left town and got a new job.

Now, in no way am I advocating that everyone should leave their spouse and go get some piercings. What I am advocating is that instead of viewing every change as a "crisis" to consider viewing it as an "awakening" instead. It is absolutely possible and responsible to have an awakening without having to turn your life upside down.

Yes, something has changed. Maybe it was up to you and maybe it wasn't. The point is that things are different and you have to change as well. Is it a crisis, to be paralyzingly fearful of and to dread, or instead is it an awakening; an opportunity to grow, adapt and thrive? The cool thing about life is that you get to make that choice for yourself.

Life can be full of lemons. It may suck, but it's true. That being said, a lemon is what you make it. If it's a *crisis* then the citrusy thing in your hand is a nauseatingly colored, hard-to-peel, sour piece of barely-edible food filled with seeds that makes you want to turn your head in repulsion. If it's an *awakening* then it's a joyously colored, interesting-to-peel, unique fruit that's chalk full of vitamin C to make you healthy! The differentiator is how you look at it.

Whatever has happened and however you're feeling, you can do this. Do it yourself or do it with help - get through it and flourish. ***Mid life awakening***

Author's note - The reason that my "failed" marriage is in quotation marks is because I believe that any relationship through which we have learned and grown is never a failed relationship. Although others may view it differently, which I fully respect, I feel as though I have not had a failed relationship in my life.

In their nature

"To gossip is like playing checkers with an evil spirit: you win occasionally but are more often trapped at your own game." - Native American Proverb

Many of us have heard a variation of the story about the scorpion and the river turtle and it goes something like this:

One day a little river turtle was gaily swimming in the river when he heard someone calling out to him. Upon swimming to the water's edge, he saw a scorpion. The scorpion said, "River Turtle, I need to get to the other side of the river - could I hop on your back and you give me a lift to the other side?" The river turtle replied, "Absolutely not! You are a scorpion - you will surely sting me and I will die!", to which the scorpion said, "But River Turtle, that wouldn't make any sense - if I sting you then you will drown and we shall both die!" The river turtle thought about it for a moment and being the kind and helpful creature that he was decided that the point made sense and allowed the scorpion to hop on his back. Half way to the other side of the river the scorpion stung the turtle. The river turtle said, "Scorpion - you have stung me! Now I can't make it to the other side and we both will die. Why did you do that?" The scorpion replied, "Because it's in my nature."

Sadly, this story can ring true in real life as well. There are good natured people who want to help others but we need to be discerning with the kind of support that we offer.

I met Steve over a decade ago and heard about his reputation as a gossip almost immediately upon taking the organization on as a new client. Steve's peers told me all about him, warning me as they wished others had warned them not to get close.

Due to the fact that Steve was a Manager, I dealt with him more than some of my other clients as I had to provide him with updates and recaps. As I got to know Steve he opened up about many things in his life; private and painful things from his past that he had never told anyone including his parents when he was younger and even his wife now. My heart ached for him and I truly wanted to help him.

When he would try to tell me personal or nasty gossip about others who had opened up to him I would council him to never repeat those things again, as it was hurtful to betray others in such an intimate way. I was touched that he trusted me and during our long talks he confided in me that due to my advice his health and personal life had tremendously improved. I really thought that he had seen the err of his ways and might change for the better.

One day a peer of his that I had a good and longstanding relationship with felt compelled to tell me something; in an effort to impress his male colleagues, Steve had publicly made incredibly inappropriate comments about me. I was devastated, I was humiliated and I was hurt. I felt stupid and taken advantage of.

One thing that I was not was surprised. I immediately thought of the story of the scorpion and the river turtle and knew that it was not Steve's fault, but mine because I should have known better. I knew the type of behavior he preferred as he had showed me time and time again with his destructive gossip. I actually smiled as I thought, "Thanks for the reminder." It was a hard pill to swallow but a wonderful gift to remind me that when people show you their true colors, you should believe them.

The saddest part is that Steve doesn't see the irony around the fact that the very actions that he takes to be part of the group are the very actions that alienate him from his peers. The gossiper has become the gossip and the person that he hurts most is himself.

Events worked out positively and my account assignment changed so that I no longer had Steve as a client. Other than a polite hello at a charity event, I never spoke to him again.

There is no anger towards Steve. I honestly wish him well and hope that one day he chooses to stop his destructive behavior. His reputation at the organization where he works has been tainted through years of passive-aggressive behavior towards his peers but it's never too late to show people that one can change. I truly hope not only for others but also for himself that he does.

For the rest of us, continue to be the good person that you are even if you get stung by a scorpion. Continue to help and encourage others as you do or want to do in the future. Be open to signs of how people choose to behave and ensure that you are careful to protect yourself and others from getting hurt. There is no need for being upset towards people who simply are who they are nor is there any reason to take their actions personally. Recognize and accept it and support them in a way that still protects you as well. And if you can find it in your heart, send them love. In their nature

It ends here

Conformity is the jailer of freedom and the enemy of growth. -John F. Kennedy

There were many things that occurred as a child and in my teens where I thought to myself, "I will never do this when I am an adult." I have not forgotten those promises. The simple yet powerful translation is, "It ends here."

Now here we are, most of us adults and every single one of us with our different scars. What a miracle that we're all at this point today; humans are such resilient, hopeful beings and each and every one of us is living proof of that. You should be so proud of yourself.

The mindful phrase, "It ends here" is not meant to be a reminder of traumatic events or an invitation to relive painful memories. It is meant to inspire you to be a new beginning, to empower you with the knowledge that you are full of strength and faith. We can't rewrite history but we can certainly change the future.

Whatever it is that needs to end, you can do this. If there has ever been a time when your body or spirit has been violated in any way either by someone else or by yourself, make the decision that it ends here. We will not pass this poison on to others nor will we allow it to live in ourselves any longer. If negativity has been a struggle for you then remember that it ends here. If there is anything in your life that has not or does not serve you and others well then make the commitment that it stops here. If you have feelings that you won't be able to do it then those thoughts need to stop here too. And if you need help, then please get it; just because it ends here doesn't mean that you have to do it alone.

There are going to be times where we may falter and if you haven't heard it enough yet you're going to hear it again; mindfulness is about compassion. Pick yourself up and start again, every moment is anew. Even if you stopped it one more time, that was still one more time than you would have versus if you hadn't ever tried. Instead of berating yourself for where you got it wrong, praise yourself for where you got it right. Then get right back on to that horse again.

Awareness is key. Be aware of the pattern that you would like to change; you are a brilliant and talented artist with a blank canvas. Although the masterpiece may be influenced by who we are, including where we have been, it's final image will be unlike anything that has ever existed before. ***It ends here***

Everybody messes up

"Our true nationality is mankind." - H. G. Wells

As you may recall, the first "C" in the three C's of mindfulness is compassion. When people don't perform the way we'd like them to, it's a fantastic opportunity to practice compassion by remembering the little ditty, "Everybody messes up." When appropriate, I like to accompany it with a little shrug.

So many of us want life to be perfect. We want our friends, family, partners, neighbors and co-workers to be perfect. Heck, we want and expect to be perfect ourselves.

Do you know what life would be like if everybody was perfect? BOR-ING!

So what harm does it do to choose understanding versus the alternative of being irritated, aggravated or angry? I don't know about you but nodding, shrugging or being mindful and quietly acknowledging that, "Everybody messes up", would be a lot better for my nervous system than getting frustrated and thinking, "What an idiot!"

When we get upset over things and start to lose our minds (well, at least the functioning of the pre-frontal cortex of our minds, as discussed in the Chapter "Override the brain"), it can be like a tornado in our heads. Tornados are never good. Things get destroyed and people get hurt.

If we can quickly catch the tornado as it starts to form and remind ourselves that everybody messes up then maybe we could stop the tornado from forming in the first place. Even a bad rainstorm is better than a tornado. If we really catch ourselves then maybe we could even find the humor in the event and turn it in to a gentle sprinkle or possibly a sweet summer day.

If somebody has messed up then it means that the event has already occurred. You can't change that. The only thing you can change now is your reaction to it.

The cashier forgot the code for your Star Fruit and has to go and ask someone else? Everyone messes up; it's not like we've never forgotten anything. Now they'll know what the code is for the next customer who needs it. You helped them learn!

The babysitter forgot bedtime and got the kids to bed way too late? OK, everybody messes up. The kids probably had a great time.

Your hair looks like it's been cut with your grandpa's original lawn mower? Yikes! The hair is already gone and the goal is now to get it fixed because no amount of negativity is going to get that hair back. If you're that upset about it then find another stylist. Everyone messes up.

What good is done by getting upset when someone has messed up?

By the way, 'someone' includes you, which means that just as we can choose kindness for others when they make a mistake or have bad judgement, we can choose it for ourselves.

It's not that we're walking around saying that nobody has to do anything right anymore. It's that people aren't perfect and once something is already done the goal should be to take a deep breath and fix the issue - not to cry, yell or pout. There may be a lot of times where this particular phrase isn't appropriate, but if you look you may find that there's a lot of times where it could be. And the sunny weather is a lot more enjoyable for ourselves and others than the tornado. ***Everybody messes up***

Loosen your grip

"When the heart weeps for what is lost, the spirit laughs for what it has found." - Sufi saying

Although I personally use all of the mindful considerations in this book, this particular one is a favorite. Trying to leave my competitive, type-A demeanour in the past, "Loosen your grip" plays repeatedly in my mind when I catch myself being controlling or inflexible.

My structured journey to mindfulness officially began in 2012 and I actively searched for different ways of looking at life with an altered awareness. At the time I was a workaholic and a control freak with the compulsive need to "fix" everything. One day during a hike a friend of mine named Ken who happens to be a stellar hockey player offered me some advice. He told me that he learned as a child in hockey lessons that people tend to grip the hockey stick tighter as they get closer and closer to the net. It's an automatic tension that arises in anticipation of making the goal. The bad news is that the tenser the hand, the less flexibility available for adjusting the shot which equates to a lowered likelihood of the desired outcome. My friendly "coach" told me that watching me get all worked up over different things in my life reminded him of himself as a youngster getting worked up to shoot the puck.

His coach was a lifesaver every time he bellowed out to little Ken, "Loosen your grip!"

Soon coach Ken could be accredited as *my* lifesaver because this became a phrase that I needed time and time again. When I would get frazzled trying to figure out how I was going to get to the grocery store on the last day of the sale for red peppers when I had already worked ten hours and had to rush home to help the kids with their homework before their bedtime, I remembered to loosen my grip. There would be another sale and the extra $2 next week wasn't going to push our financial budget over the edge. When I was putting in 60-hour work-weeks but insisting that I would be the one this year to enter the chocolate fundraising order on-line for the PAC (PTA) so it wouldn't be a burden to anyone else, I finally loosened my grip and allowed the sales representative to do it for me. When the kids would regale me with the story of their day and I felt that there was yet another opportunity for me to impart some desperately-needed knowledge on to them, I loosened my grip and stopped interrupting.

The strangest thing started happening. I found that the more I loosened my grip, even by a fraction, the easier it became to hit the target. My partner could help with picking up the groceries on the way home. My stress released and he was happy that I finally trusted him enough to "get it right." (Insert eye roll here...can you *believe* that I was so uptight about groceries?) There were a lot of Moms that *wanted* to do more to volunteer with the PAC and frankly, they were way better at some things than I was and a heck of a lot more creative! The quality of the conversations with my son and daughter improved and they listened intently when the time was right and I *did* weigh in with a suggestion. What's really cool is that the more I loosened my grip, the less tight I felt inside. It was freeing to let go just a little. Bit by bit, it helped to change my life. For anyone who is a type-A personality or could do to relax a little, this tiny mindful phrase can be a huge game-changer. Loosen your grip

Forget the guarantee

"Life is either a daring adventure or nothing at all. Security is mostly a superstition. It does not exist in nature." - Helen Keller

It was late Fall in 2003 and I was on the telephone with my Organization's Employee Assistance Program, desperately trying to book an appointment with a therapist. When I was informed that their first appointment was three weeks away, I couldn't hold it together for another second and broke down sobbing. The woman on the other end of the phone asked if I needed to speak to someone immediately and through my hiccupped sobs I squeaked out, "Yes, please."

After a brief hold Michelle came on to the line. She had a French accent and a soft demeanour. I explained that I had just come upstairs from helping my husband pack the last of his things in to the truck; the decision made that morning after years of trying was that he was leaving and we would get a divorce.

I could barely get out the words to explain my pain. We were supposed to have children together, and grandchildren. We were supposed to get old together and retire together; it was supposed to be "until death do us part." I was sad to be losing my best friend, a soul mate and all of the safety, security, love and comfort that came with the relationship. I knew that this was the right thing for us but it didn't ease the debilitating sadness and deep ache in my broken heart.

When I finished pouring out my deepest feelings and pains, Michelle surprisingly said to me in a now fierce French accent, "What do you want Alex? Do you want a guarantee?" Shocked and righteous, I tearfully and emphatically responded, "YES! YES, I DO! When you marry someone, it's supposed to be for life and you can count on it and YES I want a guarantee!" Just as quickly she dryly responded, "Then go buy something at The Bay (Department Store), because that's the only guarantee in life that you're ever going to get."

I was stunned. Her words were a proverbial slap in the face however it felt like I had been punched in the stomach. My crying stopped immediately and I could hardly breathe. It felt like time had ceased to exist as her words sunk in.

And then in one of my darkest times I did something I didn't feel I would ever do again. I laughed. With my puffy, red, tear-stained face I laughed. She was right. There were no guarantees in life - there are no guarantees about anything, ever. I felt dizzy as my entire paradigm shifted. Other than the final day of packing up the condo when it was sold, I never cried over my divorce again.

I know and respect that Michelle's technique might not work for everyone but she was brilliant and talented enough to know that it would work for me - right when I needed it the most.

Over the years I have found much comfort in those words and I have shared them with others innumerable times. They ring true for everything in this world - nothing is a guarantee. Friends who were planning on retiring with their company were rocked to the core when the organization announced sweeping layoffs and they were suddenly looking for alternate employment. Life-changing accidents happen in a flash, health is always only temporary and relationships change. Financial certainty and security can come and go, leaving families bewildered and frantically searching for the next step. Material things break, get lost or stolen or are unexpectedly gone one day.

People grow and change their minds, unforeseen pregnancies and death create a new world that can hit us between the eyes because we never saw them coming.

My girlfriend got her tubes tied after their first child. When they surprisingly got pregnant with their second her husband had a vasectomy. Their third unexpected pregnancy resulted in a healthy baby girl and just like that a family of three became a family of five. Sometimes life can be like one giant Cracker Jack box and we just never know what the prize will be.

Where I thought that accepting the fact that there are no guarantees in life would leave me feeling terrified and out of control, wanting to go hide under a blanket in a corner somewhere, the opposite was true. It actually made me feel powerful and liberated. My control is the way that I respond to situations as they occur - that's the only control we have. We are all capable human beings, each and every one of us. The answers and strength are inside of us and sometimes that strength is in the knowledge and courage to get support if we need it. As the saying goes, "That which does not kill us makes us stronger." Forget the guarantee

I just can't

"God, grant me the serenity to accept the things I cannot change, the courage to change the things I can, and the wisdom to know the difference." - Reinhold Niebuhr

There is a world of difference between "I just don't want to" and "I just can't." It personally took me over 40 years to learn this life-saving lesson.

Every day there's a list of things that we just don't want to do. Depending on any host of variables including our mood or even the weather, it may be that we just don't want to do a workout, or go to the grocery store or get out of pajamas, let alone get out of bed. Heck, sometimes some of us wish we could put a sign around our necks for the kids that reads, "I'm on strike." In speaking with hundreds of others over the years, we all have our list of things that we just don't want to do but sometimes (most times) we just need to power through and get it done. "I just don't want to" is exponentially different from "I just can't."

Years ago, at a function I was fascinated by a story from an older gentleman I had known for quite some time. I was aware that the family business was a large, multi-national manufacturing company and that his extended family was incredibly wealthy. What I hadn't known was that George worked there for years as an Executive. He told me of his expensive suits, fancy office and impressive paycheque. The issue was that George absolutely despised his job - he was a self-described "blue collar kind of guy" and hated the politics as well as being responsible to manage others. Manufacturing was not a passion for him. George described in detail that one day he went to work the same route as every other single day and when he got to the entrance for the parking lot, he couldn't take the turn to head in. His words were, "I just couldn't." So, he drove around the block again and again and again, trying to muster up the strength to take that turn but his body wouldn't let him. Spellbound and perplexed I asked him what happened next, and he simply stated with a shrug of his shoulders that he gave up, went home and quit his job that very day. I was gobsmacked. The person I was back then thought he was so incredibly irresponsible and reckless. There was no part of me that understood what he had done - given up the security, the pay, the prestige, all of it with nothing to go to. His words echoed in my brain on and off for years and every single time they did I shook my head and thought he had lost his mind.

Fast forward many years and I'm at a place where I have the opportunity to witness this event first-hand through one of my best friends.

Going to work was becoming harder and harder every single day for Anne. She was doing everything she could to continue to be that dedicated, hardworking employee yet her passion for the job was all but gone. Each day became more and more of a slog for her, something that she had never experienced in her life, but she continued to press on. I listened as she reminded herself that she was well-respected at work and that it paid well and had good benefits. It didn't matter what any of us said and she continued with the mantra that she needed to be responsible and keep pushing, pushing, pushing.

Then the day came where Anne had her own realization of "I just can't." She told me that she couldn't do one more day. Not one more second. It was hard for her to explain but she said that she literally felt as though every inch of her body and mind were screaming too loud for her to ignore, "I JUST CAN'T!" It was almost as though her soul would have preferred death over another agonizing day of trying to fake that she was happy to be at work. She finally allowed herself to recognize and acknowledge that work for her had become unbearably miserable. Her partner Elise had also seen the struggle she had been going through at work and although her recommendations for leaving in the past had been cast aside, that day Anne finally agreed.

The interesting part for Anne was that the actual decision to leave was incredibly easy. She told me that it was easier than she ever thought possible and she has never looked back. It was as though her soul had been slowly rotting inside of her and although she kept telling that horse to get up and go, it no longer could. The carrots and beatings didn't work anymore and her soul simply said no.

A huge weight had been lifted from her shoulders. Anne told everyone that her life felt like it was hers again because she finally made the decision to take control. She had never done anything like that before and prides herself on being responsible but she knew it was the right choice for her. Anne refused to continue putting a price tag on her well-being.

Watching Anne through that process inspired me so much that I decided to take a deep look inside myself and reflect on whether or not there was anything in my life that when I thought of it made me say to myself, "I just can't." There were a couple of things actually, and I made immediate changes so that I could have a more loving and nourishing relationship with myself. If we plan on living this life for a long time, we'd better do everything we can to make it a good one! The changes were hard, as one of them included extinguishing a long-time friendship that was no longer healthy, but it was worth the work. I feel more in control of my life now than I ever have. My life belongs to me, just as Anne's life belongs to her and everyone's life should belong to them.

My advice to you is not to quit your job on a whim, nor is it to go ahead and cut people out of your life willy-nilly. My advice however, is to take some quiet reflective time with that amazingly wise being inside of your body, close your eyes and ask yourself if there's anything that when you think of it makes you say to yourself, "I just can't." The answers may not be easy, but you'll know what's best for you. Love yourself enough to make the right decisions. *I just can't*

Reflection of me

"No one saves us but ourselves. No one can and no one may. We ourselves must walk the path." - Buddha

For years I have found much comfort in the saying, "The way that people treat you is a reflection of them.." This allowed me two beliefs; one was that if people were rude, mean, disloyal or any other negative action towards me, I didn't have to take it personally. It allowed me to get on with my day without letting another person's actions ruin my mood. The second indulgence that this saying gave me was someone else to point the finger at. It's much easier to point the finger at others than it is to point it at ourselves.

I still firmly assert that the way that people treat us is a reflection of them and it was only extremely recently that I fully understood the other side of that coin. Although it's completely an obvious deduction, it is a very different concept indeed to face the fact that the way that we treat others is a reflection of us.

My husband and I were best friends for many years before we starting dating. He loves me in a way that I have never been loved before. His love is healthy, unselfish, intimate and whole. Never in my life has anyone looked at me, respected me or cared for me the way that he does and it makes me feel invincible every single day. He is my teacher and I have finally learned how to fully love and be loved in this life.

The point of this incredibly personal sharing is that I have started treating others differently since learning to love myself more than I ever had before. The more love that I have for me, the more love that I have available to give to others. I like to think that I was a good person before Keith and I were a couple and that I treated others with kindness and respect but it is undeniable that the amount of compassion I have for others has increased exponentially because of the compassion I have found for myself. How can we possibly ever give something to others if we don't have it in the first place? As children we can't share our candy with the other kids if we don't have any candy ourselves, but if we have lots and lots of candy then we have so much that we can give it to every child we see. In fact, we're happy to do it because we already have more than enough. It's honestly that simple. Our embodiment is our authentic reflection.

What positive emotion would you like to give to yourself and ultimately reflect to others? Whole acceptance? Everlasting encouragement? True forgiveness? Complete adoration? How about deep compassion or unconditional love? Take a moment right now to close your eyes and imagine how different your life would look it you gave yourself that gift. Although you may be able to learn it from others you can choose to give it to yourself at any time. You could make the conscious decision to give yourself the gift of innate trust or of intrinsic value or anything else that you desire or need right this very second. The choice is in your hands; the power is yours to give. You deserve your full and complete love.

When we are frustrated with someone for anything at all, it is a representation of frustration that we have within ourselves. When we had anger or resentment towards others in the past, it was a reflection of the anger and resentment that lived inside of us. When anger does not reside in us then we have no anger to give out.

Have you ever noticed that people who are inherently angry are amazing at finding so many things during their day to make them angrier? What about that happy person that just keeps having "good" things happen to them? When I give love to others through a smile or a small gesture like holding the door for a stranger then I often find love in response. When I give out something negative, it's quite likely that I will receive the same in kind. We need to reflect who we are, and who we are needs to begin with compassion. When people look at you, what would you like them to see?

If we don't like what we are attracting then it's an opportunity for us to take a look at what we're putting out there. Give it a try - put out more of what you would like to receive and wait calmly and patiently to see what comes back. Be compassionate with others and yourself and you may find that your life starts to see the most wondrous changes. Reflection of me

Watch your thoughts

"There are always flowers for those who want to see them." - Henri Matisse

Time and again in many different ways, it has been shown that depending on what we think and how often we think about these things, we can become our thoughts.

One of the best examples I have of this phenomenon is with Sandra and her partner Jason. I met Sandra through a "Mommy and me" group shortly after my daughter was born and we became fast friends. As friends do, Sandra shared with me some of the frustrations she was dealing with in her relationship with Jason. The biggest issue by far was his negativity. Sandra explained that Jason had something terrible to say and think about everything and everyone - he didn't trust anyone's motives and looked for the bad in everything he saw. This trait was tearing Sandra apart and she didn't know how much longer she could live with the pessimism as it went against her naturally bubbly personality.

Being in the mental health industry, Sandra one day asked Jason to sit down and think about why he chose to behave in this manner. She wanted him to reflect as to where the negativity spawned from. Much to her surprise he did and the answer was telling; when Jason and his brother were little, they used to play a game while watching television. They would compete as to who could say the nastiest things about each actor, character or show they saw. He explained to Sandra that it started out as a game but that he noticed that soon it had become his habit and he just never stopped. He was entrenched in the routine of finding negativity. Jason's thoughts truly became his beliefs which became his actions and consequently his life. Regrettably his unwillingness to change his thoughts led to the demise of their relationship.

The mind is like a garden; what grows is in direct relationship to what was planted and how it was cared for. Like any responsible gardener, it's imperative that we take stock of what we're planting. Every single day it may make sense to ask ourselves, "Am I planting flowers or am I planting weeds?"

As a child I had the opportunity to witness people in my life who had very hateful, hurtful and negative thoughts, feelings and comments. Watching them interact with others, I could sense or see their faces contort as they uttered harmful statements. Very early on my elementary belief became, "Ugly thoughts make you ugly." It wasn't that these people were "ugly" to look at, it was that there was ugliness inside of them when hate was present.

My childhood-developed strategy for ensuring that the same thing didn't happen to me was that every time I had an ugly thought, I would try to immediately catch myself and counter it with a thought that was beautiful. For example, if I thought to myself, "I hate that loud dog" I would try to stop that thought and replace it truthfully with something like, "That dog sure loves his owners." The rule I had for myself was that it had to be authentic. If someone was running across the road and the thought, "What a terrible outfit" popped in to my mind then I would search enthusiastically for a replacement such as, "Her hair is so thick and healthy looking!" It may sound silly but I promise you it makes a difference. In fact, it makes many differences. Some of those impacts are that it helped me to be more aware of my thoughts, I became much better and faster at finding good traits in people, things, events and places and it was as though people could sense when I was thinking nice things about them. I love giving compliments to people because it makes me feel amazing to make other people feel good. As a result of my childhood training, this technique is now easy for me to do.

Although I fall off the wagon sometimes and can be seduced by negative thoughts that jump in to my mind, I try to remind myself of my own "game" and look for the positive. As with everything in our lives, the choice is up to us. If you're thinking that your garden could do with a little sprucing up, every moment provides us with the opportunity to plant those seeds! ***Watch your thoughts***

Authenticity and intention

"But what is happiness except the simple harmony between a man and the life he leads."

- Albert Camus

Lindy, a good friend of mine for many years came to me recently requesting some advice regarding her situation at work. She's been at the hospital she works at for almost two decades and is loyal to her employer, however there have been a lot of changes over the last few years that she hasn't agreed with. People have been hired into Senior Management positions that have limited, and in some cases, no medical training or background and the culture of the hospital has changed. She feels that in some cases political posturing has taken over patient care and is saddened to see that the strong relationships she shared with the previously close-knit group of Team Leads has morphed into a everyone-out-for-themselves mentality.

Many of the nurses under her Lead (and some that are under the Lead of her peers) have expressed their concerns as well, and Lindy has felt torn between agreeing with what's been shared versus towing the company line. Lindy senses that in order to continue to be successful in her job that she should also start playing the political game yet her goal for taking the position in the first place was to help patients and run a good, strong and respectful medical support team.

Lindy's question to me was, "How do I do what the 'new management' wants me to do even though on a deep level I disagree with it, when my only other option of voicing my displeasure and concerns may limit my career progression?"

My answer to her was simple, "Through authenticity and intention." I asked her, "Are you a good person?" (yes), "Are your actions based on the best intention?" (yes) and, "Are you happy with your actions so far of 'playing their political game'?" (no).

None of us should do things that hurt the being inside of these precious bodies that we inhabit. That is not the purpose of life. There is a physical price to pay when our actions do not align with our feelings or beliefs. One of my favorite books to illustrate this point is called *When the body says no* by Gabor Matè.

I suggested that although she needs to be careful, she continues to be authentic with her Team and others, as that is what they need most from her during these challenging times. That she goes to her Departmental and Management meetings with the intention of only doing what is the very best for the hospital she supports and the patients and staff that she cares for while still being respectful. I reminded her how important (yet understandably difficult) it is to not allow her ego to influence any of her actions; when someone has a great suggestion that she should voice her support and when someone has an idea that may not be in the best interest of staff, patients or the hospital that she should calmly and rationally explain her concerns along with a different solution. If the others choose not to accept her recommendations then she will know that she did the best she could, with the right intention.

Many of us have times and situations where we are asked to be or do something that we are not and those are the times where we have the truest opportunity to show what we stand for. It's not about being highlighted nor is it about hiding in the shadows, it's about being true to who you are and making this world a better place by allowing your truest nature to shine.

Authenticity and intention

Today's a gift

"The way to love anything is to realize that it may be lost." - Gilbert K. Chesterton

A few years ago, in a rush, I quickly approached the front doors of the building to where I was going to be meeting with some clients. As I briskly walked past an elderly gentleman struggling with his cane due to his severely handicapped legs, I felt a pang of guilt for not acknowledging him in any way. I slowed down and turned around as I said to him, "It's a beautiful day today, isn't it?" He gave me a huge toothless grin and replied loudly, looking at the sky, "God is great!"

It struck me as I continued with my day how grateful that man had been simply to be alive. He had legs that made walking a struggle and no teeth. My guess is that he had a host of other physical ailments that make life more difficult for him, yet he was so full of appreciation and love that I couldn't help but smile back at him and be inspired.

We all have struggles. We all have our daily aches and pains, whether they're physical, emotional, financial, societal, or anything else. The thing to remember is that to ride each wave in our day is to still be a part of the entire ocean; to still be a part of this experience we call life. Every day is a chance to start anew; an opportunity to be appreciative of everything there is - from dew on the morning grass to the people we love, even when we are suffering.

Sometimes life is difficult and there are times when most of us ask ourselves how we can possibly make it one more minute, let alone one more day. But we can. And we do. No matter what we're dealing with there is always an opportunity to be thankful - as long as we are still alive. We're still alive when we're here today. Today's a gift

Train your response

"Happiness is a thing to be practiced, like the violin." -John Lubbock

Recently I taught a Mindfulness Meditation class with an organizational group that I visit on a regular basis. At the end of each lesson if time permits, we open it up for discussion and dialogue. One woman brought up the fact that next to the skyscraper where they're located at is a construction area and that the sound of all of the working machinery is driving her crazy. I laughed as I nodded and told her that we're going through something very similar where I live; there is a townhouse development site just two doors down from our neighborhood and the pounding is relentless. I explained that in the beginning, when construction started that I was agitated by the nail pops occurring on our walls and bothered by all the ceaseless "noise pollution." After some consideration, I realized that it would not be healthy for me to be aggravated for the next several months during construction and that I had to change my attitude immediately. I decided to train myself to feel the vibrations of all of the noise and to imagine that the vibrations gave any tired muscles in my body a massage. Every time I became aware of the loud sounds, I would inwardly search my body for tight muscles and I would physically relax them, just as they would do if they were being worked on by a Massage Therapist. I shared with her (and the rest of the group) that the other day my husband and I were eating breakfast on the patio and I was enjoying it so much and felt so relaxed that it took me several minutes to realize that the construction was rigorously ongoing. My system had been trained and now I automatically relax without conscious thought when I hear the noise, as opposed to getting upset as I used to. My suggestion to her was to give it a try.

Although I have become a lot better at controlling and enjoying my emotions, currently there is someone in my life that results in angst every time it is required to have an interaction with her. Regrettably the situation is a necessary evil (no pun intended) and I am now working on training my response to dealing with her as one filled with compassion and well wishes. I close my eyes and wish her well even when we are not engaged so that it will be easier for me to do so when life calls for an interaction. It feels amazing to send her this positive energy; when we authentically muster it up inside of us to send to others, we can't help but be touched by our creation. My sincere wish is that she somehow feels it and that life becomes easier for her. I'm not fully there yet, but that's why it's called a Mindfulness practice; because it's continuous.

Can you imagine for one moment how different your life would be if you could train yourself to be calm, grateful, relaxed or open to situations that currently cause you grief? How would it change your daily commute in rush hour if you consciously relax your muscles with the idling of the engine instead of fuming that you're not moving anywhere or being anxious with the thought of showing up late? People use mindfulness and meditation every single day to focus on their breathing during painful or unpleasant medical procedures and the technique has been clinically proven as an effective pain-reduction method. When you're laying back in the chair at the dentist you could imagine that the light is coming from the sunshine on a tropical beach instead of from the doctor that's about to drill your tooth - which would you prefer to focus on? We can train our responses to almost anything and the outcomes can be life-altering.

It's crucially important to note here that in order for this to be a healthy experience for all involved that your "training" must be positive in nature. You will do damage to yourself and others if you train your responses to be of a negative form or connotation. Sometimes it can be challenging to find good in what is perceived as bad, but there is always a flicker of good to be found somewhere in the situation if you look hard enough. Remember that a burning inferno (of joy, love or positivity...) can begin with one spark. ***Train your response***

Disregard your roommate

"I used to think that the brain was the most wonderful organ in my body. Then I realized who was telling me this." - Emo Philips

Has there ever been a time in your life where you were frustrated with the relentless voice in your head?

Perhaps it has told you that you're not good enough, not worthy enough, that you really messed something up, you need to work harder or even that you're just plain stupid? Maybe it has a different mantra that's equally enjoyable. Are you sometimes driven crazy when thoughts won't stop pouring in, one tripping over the other, even though you're exhausted and you desperately need to sleep, concentrate or relax? Have you ever wished that it would just shut up or stop when fear, anxiety and depression insidiously sneak in and take a death-grip hold of you? What about those times when we're full of judgements for other people?

They're all rhetorical questions because there is an extreme likelihood that most of us will answer "yes" to at least one, if not all, of those questions.

The real question therefore should be, "Who the heck is the voice?" If "you" are being driven nuts by that incessant (and in most cases, negative) chatter then who is doing the chattering?

Let's affectionately consider that your mind is a roommate renting a small space in your head. The entire rest of your body is where the real you resides. We all acquired this pesky roommate in our early years and it is entirely as unique to us as we are to the world. We're paired up and that roommate is going to rent a tiny room for the duration of our time here on earth. That coupling was entirely outside of your control. The level of influence that your roommate has on your happiness and quality of life is entirely within your control.

You can allow that roommate (our crazy minds) to ruin your day or you can simply allow him or her to do their thing. Perhaps your mind is anxious, sad, angry, irresponsibly spontaneous or full of stress. Much like a harmless stranger that is near you in public and muttering crazy things to themselves as though they are in their own realm of reality, you can simply choose to allow them to exist without engaging with them in any way.

There is no need for you to actively "ignore" anything; that's a lot of unnecessary effort on your part. Just don't give the useless thoughts any attention. Consciously direct your attention elsewhere.

Your roommate exists - that's OK. Judgement, anger, resentment, frustration all need not apply to your soul; you don't want them anyway. The voice can exist on its own and although sometimes you can hear a faint whisper where at other points its positively ranting and raving, it doesn't have to affect you. If you don't like the music your mind is playing, simply decide to tune in to a different station.

Your mind has its purpose and you have yours. Your purpose is to exist in love, joy and peace; to live a grounded life full of happiness, gratitude and compassion. Its purpose is to test your conviction to living your purpose. Really, we should be thankful for such a great and motivating teacher!

Allow that voice to exist. Don't get caught up in the drama or unnecessary and destructive emotions that it is trying to seduce you with, and then merely get on with your day.

That "voice" (noise) inside of us that is not us, the one that we struggle against or fight with is just a roommate that we haven't yet established a workable and respectful relationship with. That voice is in its own reality; it doesn't need your attention. Wish it well and then get back to the *real* you - the one that chooses not to entangle with unhelpful and negative thoughts and emotions.

Just think of the life you have awaiting you when you spend your time getting to know that deeper, truer, more authentic "you." There is a new relationship for you to develop; may I introduce you to yourself? **Disregard your roommate**

Cherish the children

"Spread love everywhere you go; first of all in your own home. Give love to your children, to a wife or husband, to a next-door neighbor." - Mother Teresa

Years ago one of my clients asked me how old my children were and when I told her that my daughter was six and my son was four she replied, "Ohhhh, you're so lucky....your son's still young enough to have cat breath!" Cat breath? I had never noticed before.

That evening as I laid on the quilt beside him in his bed for night-time snuggles, I quietly cuddled up closer and smelled his little breath. Carmen was right. It was cat breath. How could I have never noticed? All of these years with my children where their beautiful little breaths had passed me by. All of those moments of cat breath were gone and I would never get them back. What else had I missed on the days where I had too much to do and the evenings when I had too much on my mind? I never wanted to miss another cat breath again. And I didn't want to miss another moment. It was an epiphany that I had to cherish the children, my children - for every moment that I could.

That realization morphed in to so many more things; enjoying the smell of my daughter's hair, the sight of sleep in their eyes when they awoke in the morning. What a gift that they were awake - I would get to be with them for another day! All of a sudden, the sound of their deafening hyena laughter when I was trying to work from home became less of an irritant and more of a treat - they were loving life and each other, and I was fortunate enough to be around for it.

I noticed that then I was adoring *all* children - crying on planes because they were scared and in need of comfort from their loving parents, covered in ice cream because they couldn't eat that delicious treat fast enough, playing peek-a-boo at the restaurant with the couple in the booth beside them. What would we do without these children? They are the next generation of world leaders, from politics to nature sustainment and possibly to parents themselves so that the human race will continue. We are so lucky for every moment that we have with them.

Cherish their silliness. Cherish their hugs and needs for affection. Cherish their desire for support on bad days and need for independence on their good days. Cherish these little beings that have been brought into the world.

Cherish them even when it's not easy to do. **Cherish the children**

Always choose love

"The way is not in the sky. The way is in the heart." - Buddha

This journey of life is full of choices and mindfulness is the greatest tool to help us take the richest actions possible. From the choice of the very first thought we think when we wake up in the morning to the very last thought at night as we drift off to sleep, we get the human right to decide for ourselves what those thoughts will be. No matter what is going on in your life, nobody can ever take that right away from you. You yourself make the decision to create thoughts of love, or thoughts of anything other than love.

Years ago, after a weekend retreat, I was inspired to take the challenge posed by the main speaker to wake up every morning and have my first thought be, "How can I best serve the world today?" It took a lot of practice and there were many mornings that my first thought was regarding how I slept or what was on the to-do list for the day. However, as soon as I realized that I forgot to ask myself, "How can I best serve the world today?", I would ask it at that very moment. Sometimes I had already hopped out of bed so I would clamber back in, lie down and ask myself the question. I would ask it until I actually meant it. Some mornings the question would repeat over and over again in my head because I was going through the *motion* without including the *emotion*, so I would ask it until I felt it in my heart.

Some mornings I would answer that question. I would decide to make the effort to be kind in traffic, open doors for others, give my everything to every person I would meet that day or some other version of making the world a better place. Sometimes there was no answer, just an intention - to make a positive contribution to this world on this day. It took weeks of practice and eventually I had trained myself to have that be the very first thought I had when I woke up. It became exciting to wake up because I couldn't wait to have that thought.

Every single choice is either love, or something other than love. Those are our two options. When you have the opportunity to choose anger, choose love instead. When given the chance to feel pain, choose to feel love. When unnecessary competition or negativity rears its ugly head, mindfully choose love as your alternative. Choosing love is choosing to change your entire life.

I know that love may not be the first thing we think of when we get a paper cut, receive bad news, get deceived and betrayed, have our feelings hurt or any other myriad of unpleasant experiences in this life. But it could be. The choice is yours. And if you missed that opportunity at the beginning, it's never too late to choose love. Choose it now, choose it later - it's there inside of you if you look deep enough.

We are all human. We all make mistakes. We all have times in our lives when choosing love is our very last desire. That's OK - we're not here to be perfect, we're here to be better than we were a moment ago. One more loving thought or loving action in this world is still one more loving thought or loving action in this world. And it's one less choice that doesn't include love.

If this isn't your deal right off the bat, that's OK too - start small. Try it out with your family, your friends or your pets. First and foremost, try it out with yourself. You don't even have to tell anyone you're doing it. You can just send the thought, "May you be well" to someone, share a smile or elect not to engage with the negative behavior of others. Love comes in all shapes and sizes and you don't have to post a big neon sign above your head that says, "I'm choosing love!"

When you see someone you care about, feel love. When someone is in need, give love. When someone is hurting, send love. When someone is behaving in a non-loving manner, think loving thoughts. Like everything in life, it gets easier the more we practice. Always choose love

Time to celebrate

"In the midst of chaos, there is also opportunity." - Sun Tzu

Although many of us believe that life is supposed to be fun, we live our lives thinking that life is supposed to be fun but not actually living a fun life. It seems that most people at one point or another, either in childhood or as adults find that life is actually hard. Sometimes really hard.

That's not the way it's supposed to be. We get one life while we're here and whether you believe in a higher power or not, I don't think the mandate was, "Go down there, work hard and don't have any fun!" More so, if there's a higher power, don't you think she/he/it/they would have a heartfelt wish of, "Make a difference and enjoy every second!" I like to think it would be the latter versus the former.

The question often asked is, "How can I have fun when life is so hard?", and that's a fair question. The answer is, do it in the cracks. In between the times when life is hard, find the spot that you can celebrate. Sometimes the spots are so small that you might miss them if you sneeze, but they get easier to find the more you look for them. Celebrate that you woke up today, celebrate that you had a bed to sleep in or that you had an awesome sleep. You can do that simply by showing affection towards a loved one (including a pet or beloved stuffy). Celebrate at breakfast that you get to have a healthy meal by toasting with water, juice or milk. Celebrate having a reliable car by belting out your favorite song on the way to work. Celebrate being at work by having an appropriate little desk side joke calendar and turn the page. Celebrate that you have a great friend by going out for a walk together. Celebrate that you are in the mood to celebrate by smiling at a stranger. The important thing is to acknowledge every single celebration that you have. Acknowledge that you are celebrating - tell yourself and if it's appropriate, tell others as well. The amazing thing about celebrating everything is that very quickly you realize how much you have to celebrate. It feels good. And it's fun. And all of a sudden life isn't as hard as it used to be.

There are times when you don't *want* to celebrate. In fact, you're mad, or sad or irritated or frustrated or any other group of emotions that come naturally when things don't go the way we want them to go. That's OK. Try to celebrate anyway. Your baby will not stop crying and you are going crazy - make sure that they're safe, go and sit in another room for a moment and celebrate with a deep loving breath that you are *wise enough* to know when you've *had enough*. If you've made a poor decision and said or did something that you should not have done, take a moment to yourself to celebrate that you're mindful enough to admit and fix a mistake and then go do it. If someone has hurt you so badly and you want to cry, find someone you trust that can help you and celebrate their support by leaning on them.

There are times when life is hard. You've been there, I've been there, we've all been there. Many of us are there now. There is no denying that things get difficult and that's why it's more important than ever to squeeze joy in to every possible moment that we can. You'll likely find that the more you celebrate, the more reasons you find to continue the party. Time to celebrate

Check your baggage

"How blessed are some people, whose lives have no fears, no dreads, to whom sleep is a blessing that comes nightly, and brings nothing but sweet dreams." - Bram Stoker

As incidences and levels of mental illness continue to climb higher every single day, employers are alarmed and in some cases are at a loss as to what they can do to curtail it. More importantly, the people who are suffering are in need of help. Most of us have had times where there is so much going on that we feel that we can't possibly focus on the task at hand because everything else is too overwhelming. That's where checking your baggage can come in very handy.

In my mid to late twenties before children and with my career still climbing, I was at one of the spots where life is supposed to be the sweetest. Try telling that to my mind because it wouldn't stop. I was stressed about everything. I was stressed about my career, stressed about my home situation, about finances, issues with extended family, health and everything else that plagues so many of us. It was affecting my sleep, my personal situation and my professional life. To be perfectly honest with you, it sucked.

One evening when yet again I exhaustedly awaited my ever-elusive sleep, I imaged that I was scooping up all of my "baggage" together. I imaged suitcase after suitcase, each one of them was another item on my list of worries. I packed enough bags to fill an entire bus in my mind and then I shared what I had done with my partner and asked him to get rid of them for me. He thought for a moment and said that he had sent the (driverless) bus in to outer space and that due to air pressure issues the bus had exploded and all of its contents had been annihilated.

It was so silly. And I felt a million times better. I felt lighter. We both laughed and then we both went to sleep.

It's a technique that I still use to this day and have introduced to my children as well.

If you're not ready to imagine your baggage being burned in a safely supervised bonfire, blown up in a secure area or eaten by bugs that love devouring suitcases then just set them outside the door. Imagine packing them all up with your worries and stresses and then leave them in the backyard for when you're ready to go back to them. Leave them with the baggage check when it's time for a good sleep, a family meal or a night off. Leave them with the baggage check when it's time for work. Who knows, maybe if you leave them there long enough, you'll decide that you don't actually need some of them back. ***Check your baggage***

Repack your bags

"Progress is impossible without change; and those who cannot change their minds cannot change anything." - George Bernard Shaw

Different from "Check your baggage" where you imagine your bags being destroyed or held temporarily, repacking your bags is about mindfully taking a good look inside at the contents of the baggage you're already carrying and recognizing what can be replaced or discarded.

Over the years during hundreds of conversations with clients, the heaviest baggage that people were carrying was the baggage full of negative emotions.

Think about it - how heavy is hate? When you imagine something or someone that you love you feel light, sparkly or floaty. You want to smile and it feels warm and good. On the other hand, when you think of something that you immensely dislike it makes you feel heavy, angry or sad. You want to cry or seek revenge and it makes you feel bad. Negative emotions are literally exhausting to carry. Consider the energy it takes to have positive emotions such as love, joy, forgiveness, kindness and compassion. Those emotions actually *give* you energy. Now think about the emotions of hurt, betrayal, bitterness and envy. Those emotions *consume* your energy.

Just as junk food tastes great when we eat it, we feel terrible when we have too much. It's called "junk food" for a reason - it's junk, and we don't want to put junk in to our bodies. Don't get me wrong, I know it's bad for me but I don't ever foresee myself pulling away from my desire for chocolate-covered jujubes. We're all seduced by negative emotions sometimes, just as those chocolate little delights call my name on Friday nights. So, indulge if you must, but then know when to stop indulging. Walk away from junk food and negative emotions when you can. They both poison our bodies and make us heavy if we take in too much.

What are you carrying with you that needs to be repacked? Are you carrying hurt, anger, betrayal or hate from something in your past? Do you really need to carry it with you in your bag? The bag gets heavy and you have far better things to spend your energy on; positive things.

So many of our bags can be repacked with forgiveness. If we've done something that we regret then don't do it again, fix what you can and practice self-compassion by forgiving yourself. Leave the regret behind. If someone else has done something that we wish they hadn't, then forgive them for making a bad decision. Forgiveness is not the same as reconciliation - you don't have to let that person back in to your life again if you choose not to. It just means that you don't carry the negative emotions that you no longer need. It'll make your load a lot lighter. **Repack your bags**

Acknowledge your emotions

"The rose and the thorn, and sorrow and gladness are linked together." - Saadi

A key aspect of mindfulness is to authentically acknowledge and experience what is happening in the moment that it occurs. Nurturing our emotions is an excellent way to practice mindfulness. It's very simple, and is so often the case with things that are simple, it is not always easy.

Consider your emotions as a crying baby. Generally, our first instinct upon hearing our baby cry is to rush over and comfort them. We might lift him or her up, hold them snuggly yet gently in our arms, sway them back and forth and murmur comforting words to them. When the baby is soothed and feels safe, secure and acknowledged he or she will stop crying. This may take only a few moments, or if our baby is colicky it may feel like the process takes eons.

Our emotions cry and scream out to us because they are needing to be acknowledged and comforted. Often what we do instead of taking our time to explore their needs is we push them aside or judge them for calling out to us in the first place. Neither of those things would nurture a baby and neither of those things will nurture your emotions. Indeed, you may be able to get on with your day but where did those emotions go? Just like the crying baby that you no longer hear, they're not gone, you just left them in the other room.

Earlier this year my husband and I went on a trip together that included a silent retreat. (Yes, I have heard the jokes that people who pay to have others not speak to them or look at them should just hang out with their enemies and save the money.) Being in silence is a tricky thing and often brings rise to many thoughts and emotions - it's like walking in to a room and being faced with all the crying babies you've left over the years.

During the days after the retreat as we navigated through the beautiful States of America Keith told me that one of the issues that came up for him was that I had previously dated a man that didn't treat me very well. As we explored his thoughts it came out at the very root that he was angry at me for allowing that to happen and for not walking away sooner. It was such a beautiful and brave emotional epiphany for him to have. As most of us do, he then immediately started judging his emotion. He told me that it was ridiculous for him to feel that way; that we hadn't even known each other at the time and it didn't make any sense at all for him to be mad. "Ridiculous." How can the way that we *feel* be ridiculous? Just as the way our bodies *look* can't be ridiculous the way our bodies and emotions *feel* is not ridiculous. It is what it is, they are what they are, as simple as that. And it all needs to be acknowledged. Now, how we *behave* because of the emotion may be ridiculous if we do something hasty and irresponsible but how we *feel* with the emotion is never to be dismissed. Keith allowed himself to sit with the anger for some time - no longer judging it and just allowing it to be there. "It is what it is" gradually became, "it was what it was." He was able to move past the emotion while in no way condoning the way that I had been treated by another human being or judging me for not leaving earlier. He had comforted the baby just by holding it without telling it to stop crying.

We hold back our emotions every single day. If we didn't, we would be unable to perform in society. There are times when it is safe and responsible to acknowledge our emotions and other times when we truly do need to let them lay for some time until we can get back to them. That's OK. Take your time and get support if you need it. As terrible as it is to face, the reason that there is a name for Shaken Baby Syndrome is because it is something that happens when we're not in the frame of mind to offer care by ourselves. Take a breath. Know what you can handle, know what you can't, and get help if you need it.

Our emotions are an extension of ourselves whether we like them or not. The only way that we can truly, deeply love others is when we truly, deeply love ourselves - every aspect of ourselves. Be kind, compassionate and judgeless. You deserve it. *Acknowledge your emotions*

Discard useless lessons

"If you want to succeed you should strike out on new paths, rather than travel the worn paths of accepted success." - John Rockefeller

We are all taught lessons from the very moment we are born. From learning that someone lovingly responds to our cries (or not) as a baby to how to properly operate the washing machine when we move out, we learn lessons. Some of our lessons are extremely useful and help us to live more happy, healthy and productive lives. Other lessons work against us where they can at times cause us to be unhappy, unhealthy and unproductive. One of the many interesting things about learning lessons throughout our lives is that very few of us are ever taught the vital lesson to discard useless lessons.

During my first Maternity Leave I enrolled in a program to operate an at-home licensed Day Care. I learned many lessons from that course and one of the most important was that I learned I'm not cut out to operate an at-home licensed Day Care. Anyway, I digress.

Another lesson was regarding proper and healthy eating habits for children. There was a long list of "never-to-do's" with regards to introducing kids to their life-long relationship with food and eating. As I sat there scanning the list in stunned silence, the full impact of what I was reading rocked my world. Literally each and every single item listed there as what "never-to-do" was how I had grown up in my family learning how to eat. Yes, I was taught to eat every single thing on my plate whether I was hungry or not. Yes, if I did not finish my meal by the time the meal was over, I had to sit there by myself until I became "hungry enough" to finish eating. Yes, if I didn't finish my dinner from the evening before and my parents were tired of supervising me to ensure that I didn't throw it out or feed it to the dog, I had to eat my leftover dinner the next morning for breakfast. Yes, I had my meals dished up for me by someone else who didn't know how my appetite was doing that day. Yes, food was used as a reward. Yes, meals were withheld as punishment. As a child, the dinner table was a breeding ground for abuse, not just with food but also with the threat and follow-through of being forcefully hit with utensils if manners were accidently neglected. Very often, eating was a negative experience. My lessons around eating had been taught to me by ill-equipped teachers.

Suddenly I had clarity around my entire life-long relationship with food. I understood more clearly why I had been anorexic as a teen; in my unhealthy mind I was choosing to finally take control of my life by refusing to be controlled by food and hunger. I understood why in adulthood I would finish a meal or dessert even if I didn't particularly enjoy it, why I would sit after dinner was over to finish every morsel of food on my plate and why I fought so hard with my body image.

Thankfully as I grew healthier, I started questioning everything in my life, including my lessons. I am so tremendously grateful that I love myself and others enough to have chosen to leave those harmful lessons behind.

Mindfulness offers us the opportunity to quietly reflect on lessons that we have learned throughout our lives and to question whether or not they are healthy or necessary. Some lessons may have been useful in the past and are no longer needed while other lessons may have been flawed from the very beginning. We are at a point now where we can take some breaths and look at our actions and beliefs with curiosity and without judgement around whether or not they are something that serves us well. Remember to proceed with compassion. This may be something that you do alone, with a partner, friend or family member or with professional help and can be an incredibly eye-opening and liberating experience. Discard useless lessons

Do no harm

"The best way to find yourself, is to lose yourself in the service of others." - Ghandi

This phrase is two-fold and both parts are equally as important. The first side of this coin is to do no harm to others in any way. This would include not only in physically not harming others but emotionally as well. It is possible for us to harm others with our actions, our thoughts and our words.

Most of us have said something very nasty to or about another person. The vindictiveness behind the action can be intoxicating and difficult to resist. It can feel artificially satisfying and gratifying to know that we've stung someone deeply.

What I've noticed more as I continue my Mindfulness Practice is that as soon as the initial comment or action is over, if I turn the rock of satisfaction to the other side, I find pain. The pain is mine. Both because of the pain that I've inflicted on the other person and because it truly does hurt us to hurt others. For lack of a better word, when the event is over, I feel "bad." The sweetness of vengeance suddenly leaves a bitter taste in our mouths.

Over the years as I reduced the things with mal-intent that I did to others, I came to recognize that even mean thoughts had the same effect on me. In my crazy mind I would think of an insult directed towards them or "karma catching up to them" and although initially it would feel justified, more and more it would leave me feeling empty afterwards. The hatred that was in my actions towards them, spoken or unspoken, left bits of that hatred inside of me. My anger was making me sick, both physically and emotionally. It hid there for so long, so many years, until I became more in touch with myself and how I actually felt when I did things with an ugly intention. It has come to the point now that the moment my intentions are impure, most of the time I feel the ugliness immediately and take correction action. I'm extremely thankful for that as it reminds me what of I don't want to be or do.

There is no good that comes from harming others. Think back to your childhood. Are you happy with the mean things that you said and/or did to another child? Do you still feel that same satisfaction that you did so long ago, or is there regret inside of you now that you can guess the implications that your words or actions had on another human being?

Other than protecting oneself or another being against imminent danger, there is no need to harm another. We were not placed on this earth to go around hurting others with our words or actions. It's so easy for us to gossip about others, use hand gestures out the window to someone we feel is driving badly, say terrible things about people that we don't want others to like and to hurt others directly with our words or actions. When we look back at these events with clarity, it's very rare that we feel good about what we have done or how we have behaved. How has it affected that other person? We've all seen others who have had their reputations destroyed by gossip, most of us have been the recipient of a hand gesture that we didn't feel was necessary and we carried it with us for at least part of the day. What is the life-long impact on the child that no longer trusts his/her other parent because of something that was said?

Through the years I have seen and heard about so many ways that people were affected by things that others did to harm them. Adults that are unable to trust others because of sexual abuse, people who suffer from tremendously painful auto-immune disorders that they can directly relate to an episode where someone purposely set to cause them emotional pain. I would like to think that if most of us knew the long-term implications of our hurtful intentions to others that we would never choose to take that path. Hurt people hurt people. Let's stop hurting people so that people stop hurting people. Allow the buck to stop with you.

That brings us to the second side of the coin. "Do no harm" is intended to stop us from harming *anyone* and "anyone" includes ourselves. So many of us try not to harm others but don't give it a second thought when we harm ourselves. We cause harm to us when we have negative thoughts or actions about or towards ourselves. We cause harm to us when we make decisions that are not nourishing. We cause harm to ourselves when we don't care for ourselves the way we should. This is anything from not pausing when we're tired to self-mutilation and everything in between. So many of us cause self-harm by overeating too often, drinking/smoking/using drugs past a level that we know we can handle or not getting help when we need it. So many people cause devastating self-harm with their hateful thoughts to and about themselves. “Do no harm” means do no harm to anyone, including you.

If you want to change your world, do no harm. If you want to improve your relationships, do no harm. If you want to live a happier life, do no harm. If you want to love yourself, do no harm. Start today - start now.

Do no harm

Life's a mirror

"The world is like a mirror. See? Smile, and your friends smile back." - Zen saying

Sometimes life is a one-way mirror and sometimes life is a two-way mirror. The difference between which is which is based on the behavior that you are giving measured against the behavior that you are receiving. The best instrument to use for deciphering between the two is mindfulness.

When life is a two-way mirror it is exhibited by two or more people behaving in a similar way. Often anger begets anger, egotism begets egotism and joy begets joy. If you are having a heated conversation with someone, take a look in the mirror. Are you reflecting what they're reflecting or are they reflecting what you're reflecting?

It's a moot point. By the time that you calm down enough to ask yourself that question, you can then choose to be mindful enough to change your reflection. If there is unnecessary anger then perhaps the right choice would be to reflect compassion and if there is unnecessary sadness then you may choose to reflect hope or levity. You get the point. Take a look at what you're receiving and if it's what you're also giving then it may be time to change your reflection.

Conversely, when life is a one-way mirror it is exposed to us through the awareness that we are receiving something different than what we are reflecting.

While enjoying breakfast one morning with a self-identified Empath who can sense other people's emotions, Sheila confided in me that the most difficult part of being an Empath was that she was a mirror for everyone that she spoke with. I was intrigued and prompted her for more details. She said that when she's having a conversation with someone, they will often become angry for no reason, sad for no reason, confused or elated - any host of emotions. She said that they tend to then say to her, "I don't know what it is about you that makes me like this!", and will often be displeased with their behavior and blame her for "bringing it out in them." Sheila explained that when she speaks to people, she just holds their energy without judgement and then they see what they're reflecting.

Think back to the time when a perfect stranger was raging about something seemingly insignificant in public. That person was reflecting who they were at that moment. What about the episode where a perfect stranger smiled so deeply at you that it warmed you to your core? You didn't smile at them first, so it wasn't a reflection *from* you, it was a reflection *to* you, which means that it was a reflection of them at that time.

Our reflections change from moment to moment depending on what is, or has occurred in our life. If you don't believe me, step in front of the mirror and take a look for yourself. Did your eyes just blink? Did your chest move with your inhale? Can you see the vein in your neck pulsing? Your reflection changes constantly.

When someone is behaving in any sort of a way towards you - positively or negatively - consider the reflection. Is it a two-way mirror and you're both behaving in the same way? If so, is there anything that you would like to change about your reflection? If someone is behaving in a manner that is different from yours then perhaps you are acting as a one-way mirror for them and what you are viewing has absolutely nothing to do with you, as their behavior is just a reflection of them and how they're feeling in that moment.

Now, that doesn't mean that we can't take a look at our actions and inquire mindfully to see if there was something that we did that could have facilitated that reaction in them. Perhaps they took something we said or did in a way that it was not intended and then we can adjust our behavior to reflect the true intention.

It does mean, however that we do not need to own their behavior nor do we need to take it personally. Even though they may try to make us think otherwise, it's possible that their behavior has absolutely nothing to do with us.

We also need to remember that just as our reflection will change with our mood, daily events, life circumstances, level of sleep and countless other variables that others are allowed that same understanding. A reflection in one moment is not necessarily a reflection of that person overall. Be calm, be mindful, be compassionate. ***Life's a mirror***

Used to be

"Reflect upon your present blessings, of which every man has plenty; not on your past misfortunes of which all men have some." - Charles Dickens

A long-time girlfriend of mine would often speak to me about various issues she was having with her husband Tom. Although they had a wonderful life together, two happy and healthy boys, a beautiful home and successful careers Tom was often stuck in the past with the way that things "used to be."

Pam was concerned with Tom's out of control spending and whenever she brought it up, he would become irrational and carry-on about how he had grown up poor, that the mere word "budget" brought up ugly memories from his childhood and she had to stop the conversation immediately. When he had too much to drink, he would miser about memories from his past including Christmases where he didn't get the gifts that he wanted and other "used to be" memories. Tom knew that he had some issues to work through and would often say to Pam, "How come my brothers can get past all of this and I can't?" Although Pam had provided Tom with a life of luxury and urged him to seek professional help he refused. Years later when they divorced there were limited assets to split due to his overspending, compensating for what his life used to be. He was once again feeling abandoned and deprived yet what he couldn't see was that he had helped to bring it on himself by refusing to let go of the past.

So many of us are stuck in a used-to-be mentality. Some of us lament over how wonderful life used to be while others find it challenging to move past how hard life used to be.

The operative word in this phrase is "used" which depicts the old way or in the past. Although the past certainly can and often does shape us in to being who we are, it is no longer *where* we are. Where we are is in the here-and-now; the past is a memory and the present is the only place where we can make new memories. Living where things "used to be" is like believing that your life is a TV show. It's not reality and only lives in our minds - we're "watching" memories believing them to be part of our current life. We can't touch it, smell it, taste, hear or see it. We can however, touch the present. Currently, you are touching this book. We can smell the air right now, hear the sounds from outside right now - change the future right now.

Be compassionate with yourself. If what your life used to be like was so much more fun then add more fun in to your life now...in the present. If your life used to be challenging then get help to make it easier for you to reconcile and live a happier life now...in the present.

Try to be grateful for the lessons that you've learned and that you have a chance to continue your life to this moment. This very moment where there is no such thing as "used to be" because this moment is "what is." Take action to ensure that the rest of your days are happier. Let go of what used to be so that you can fully extend your arms and embrace what is. **Used to be**

Give a moment

"All happiness or unhappiness solely depends upon the quality of the object to which we are attached by love." - Baruch Spinoza

Mindfulness is about being present in this very moment, and the potential for compassion is in every single moment that we give to another person.

It is completely possible for a teacher to take a moment and answer a child's question but that doesn't necessarily mean that the child received a moment from that teacher. The "moment" is that sweet spot where awareness and presence meet and are offered to the other party. The magic of the moment was in the authenticity of the smile, the tenderness of the words or the visible compassion in the eye contact. Those things were the moment, not the answer itself.

Consider the romantic comment, "We shared a moment." The statement implies that there was something enchanting and delightful present, almost like a special secret. How often do you "share a moment" with your partner, family members or friends? In order to share anything in life, you have to actually give it to the other party. Giving a moment includes your undivided attention, as if that person, place or job is the only thing that exists. You know the feeling; the one where you notice the warmness in someone's eyes, the beauty that surrounds you or the awe at how focused you can be on one single task.

The good thing about giving a moment is that it only takes a moment to do. The other good thing about giving a moment is that if you're lucky, it can last a lifetime. Give a moment

Everyone wants happiness

"Our minds are as different as our faces: we are all traveling to one destination - happiness; but few are going by the same road." - *Colton*

One thing that is universally agreed to as humans is that we all want to be happy.

That's where the easy part ends, because what makes people happy varies from one person to the next. Some people find happiness in health, others in wealth or power. Friends make many people happy as does spending time alone or in nature. Happiness may be described as a night on the town or an evening in the bathtub. Maybe it's in goodwill for all or in a favorite team winning the Superbowl. The point is that we all want happiness but that the definition of happiness changes depending on whom you ask.

If we're all equally human, then does that mean that anyone should be allotted more or less happiness than anyone else? Is my happiness more important that yours, or his happiness less important than hers? Of course not, we're all allowed to be equally happy. Happiness is our birthright.

Here comes the tricky part - what if what makes one person happy comes at the expense of what makes another person happy? When three hundred people are in a bike race it is likely that one person will be very happy to win and almost as likely that two hundred and ninety-nine people will not be happy to lose. Does that mean that those two hundred and ninety-nine people are losers? No, it means that they simply didn't get what they thought would make them happy that day.

When we are mindful, we are able to recognize how similar we all are and that can offer solace during challenging times. That person that won the race just wants to be happy. The others want to be happy as well and mindfulness will help them to recognize and accept that everyone wanted to win and only one person would get what they wanted. The same is true with people who cut you off in traffic. It may not be right, but that person's deepest desire is to be happy and maybe happiness to them is being home sooner or being a bully to other drivers. It doesn't change the fact that the similarity between you and the person who cut you off still exists. You both want to be happy and you both have different definitions of what happiness is.

My offer to you is this - when you are challenged with the behavior of another person or in the outcome of an event, try remembering that one of the ways that humans are all connected is by our wish to be happy. We are not here to judge or condone what makes another person happy. Accepting that desiring happiness is a universally human trait can sometimes make it easier to have more compassion with others and ourselves when we don't always get what we want. *Everyone wants happiness*

Try less anger

"Silence is true wisdom's best reply." - Euripides

Anger is an insidious disease, and one of the most contagious that plagues our planet today. It doesn't affect only the person who is angry, but it affects anyone who is the recipient of the anger and startlingly, anyone who even witnesses or hears about the angry event. When we are involved in anger in any way our heart rate goes up, our adrenal glands start pumping adrenaline through our body and the chemical cortisol is released in to our blood stream. Nothing good happens in our body when we are involved with anger in any way. The more entwined the relationship with anger the more of a beating that our bodies take.

Can you imagine if anger caused immediate death? The planet as we know it would cease to exist within days. People would be struck with "anger death" in such trivial matters as traffic and parking lots, retail stores and restaurants, in waiting lines or telephone queues - the list is endless. Of course, anger wouldn't be worth immediate death, but then why do we allow these things to affect our health at all? What if instead we made the choice to *not* be angry?

I understand that sometimes we all have anger and frankly so many of us are in, or have had, angry times in our lives. This is not a finger-pointing, judgement-making demand of, "Let anger go!" but an open-handed offering to simply just, "Try less anger." There is no judgement here as I too have had angry times in my life. In my twenties I was an especially angry driver and I wish that I could take back all of those words and thoughts directed at other people. They didn't improve my day and they certainly didn't improve the day of anyone else that was a party to the event. I didn't get to where I wanted to go any faster and it took such a toll on my emotional well being, not to mention the emotions of others. A difficult and painful question to ask is, "What were the impacts of that anger?" Did these people then go and spread the anger that I had spewed at them to other drivers, their co-workers, their spouses or even their children? So often if we knew the impact of our choices we would choose differently. Making the conscious choice to not be angry would have made my life and the lives of others a much more peaceful place to be.

It's hard at this point for most people reading this to not say, "But wait a minute - sometimes people are idiots and deserve for me to get angry at them!" Yes, I will agree that there are a lot of times when we feel as though we are in the right and they are in the wrong and we would like them to "smarten up." Is getting angry at them really going to change them? Will they look at you and say, "You're so right - I was so wrong and I deeply apologize"? Likely not. And what have you done to your health and theirs in the meantime? If you look back at as many times as you can remember of angry episodes, do you still feel vindicated and truly believe that you couldn't have handled the situation any better? I certainly don't. In fact, there are very few times in my life that I reflect upon and say, "Yup - so glad I got angry about that!"

Leah, an old teacher of mine once told a group of us that she was so angry as a teenager and in her early twenties that she used to run in to crowds of people and just start punching. She said it didn't matter who it was, she just wanted to hit as many people as possible. Leah stated that she had so much anger inside of her that she just wanted to lash out and hurt people. Her regret and pain was palpable. I couldn't help but wonder if one of the reasons that she became a teacher was to try to right some of the wrong that she feels she has done in this world. Those actions will be with her and all of the people she affected until the day she dies. I don't want that kind of regret; do you?

We will all get angry. Unless you're a Saint it's going to happen to the best of us. Be aware when it occurs and see what happens when you allow that anger to lift just a little bit. See what happens if you decide that for this one event, you'll let the anger loosen or even decide to surpass it and let it go. Just because you were angry a moment ago doesn't mean that you need to continue to be angry in this moment. Trust me when I say that with practice it becomes much easier to just drop your anger in a heartbeat. You may find that it feels so good to your own health that it's something you may try again...and again. The journey of no longer being an angry person starts with the step of having just a little less anger. Try it for yourself. Try it for your family. Try it for all the other beings on this planet. Try less anger

This is life

"Half our life is spent trying to find something to do with the time we have rushed through life trying to save." - Will Rogers

Surveying the line ups at the grocery store I decided to go with the shorter line and the woman with the large buggy of items versus the longer line-up with several people who had only a few items to purchase. I deduced that one person with lots of items would be more effective in pushing through than lots of people with less items who would all have to pay and bag individually. Not surprisingly, my deduction was wrong.

As I stood in the line up now watching all of the people in the other lane pay and leave, pay and leave, I could feel the impatience welling up inside of me. Knowing that my head was going to be an ugly place in a very short amount of time, I took a deep breath and relaxed. "This is life", I thought. It wasn't even a, "you win some, you lose some" situation, it was just life. Sometimes it works out the way you planned and sometimes it doesn't. As the cashier requested a price check over the intercom, I took a look at the magazine covers and made small talk with the nice people behind me. Then I took a look around and learned that they have a place to check lottery tickets and realized that it will save me from having to go halfway through the mall to the lottery booth. Those extra minutes in the line up will save me time over the long run but it doesn't matter - waiting in a line up is just life.

There's an Adam Sandler movie where his character is provided with a remote control for his life and he realizes that he can fast-forward through boring events and fights with his spouse. He's so excited and takes advantage of the amazing tool until it dawns on him that his entire life has passed him by. Life *is* those mundane tasks. Life is those fights, those never-ending recitals and boring staff meetings. Life is hidden in those silly moments and those "wastes of time." Just look for the gems - the sights, the sounds, the feelings and especially, the learnings and the love. I had an opportunity to practice being more patient and mindful in that lineup. What a gift! Instead of wishing these events away, try making it a goal to gleam everything you can in the moments that you have. This is life

The bright side

"The most certain sign of wisdom is cheerfulness." - Michel de Montaigne

Our last family vacation was one that we had been looking forward to for a year. All of the planning and finally the time had arrived! We went up to Whistler, British Columbia for our week of hiking, biking, tennis, swimming and more only to find that the winds had changed and the entire city was covered in wildfire smoke that was so dense we literally could no longer see the epic mountain views from our hotel windows. It was a letdown to say the least.

Now, I realize both at the time and now that I was being incredibly selfish and that much larger things were at play than just our family vacation. Countless brave and selfless firefighters and volunteers were putting themselves in danger every single minute of every single day to put out those fires, risking their lives to save the lives and livelihood of perfect strangers. They are all heroes.

Regrettably at times, it is still difficult for us to get past human emotions and at that time, the human emotion I felt was disappointment.

After that first day of feeling sorry for myself, I told my husband that I was sick of my attitude and asked him if we could play a game before we went to bed. Being the good sport that he is and knowing that I can be a bit kooky sometimes he agreed. The name of the game was "The bright side" and we took turns talking about all of the bright sides to being trapped inside on our vacation due to smoke (that was actually coming in the closed windows and doors and caused one of the neighboring units to have their smoke detector go off).

The reasons were plentiful – on the bright side we had remembered to pack a whole bunch of board games and a deck of cards, we had access to free DVD's and a DVD player, we were able to check out books at the local library and although the room was small, we considered it "quaint." The game continued and there were tons of points on the bright side just as there always are in any situation.

Sometimes one of the best bright sides is that one day you'll look back and have a story to tell.

That conversation changed the entire trip; it was exactly what we both needed to get out of our funk and accept what was. And to make the best of it. Our son in particular told us before we went home that it had been the week of a lifetime. Now that's a lot better than all of us going home feeling like it was a terrible and wasted vacation.

What happens is what happens and sometimes it truly is out of our control. How we respond to what happens is completely up to us. Make your choice. **The bright side**

Take a moment

"To a mind that is still, the whole universe surrenders." - Chuang-Tzu

The moment that it takes for you to read this sentence will never happen again. Neither will this one.

Our entire lives are made up of nothing more than moments, one after the other until the day we take our last breath. Being mindful makes them count.

Take a moment right now to be fully present. Feel the physical sensation of your body being supported by the seat or your feet being supported by the ground beneath you. Take a deep breath and fully surrender to life for just a moment - allow your shoulders to drop, your stomach muscles to soften and any stress or tension you're holding to just melt away. If your worries are knocking at the door of your head, let them know they can wait and you'll be with them in a different moment.

Take a moment to think of something that you're incredibly grateful for including the fact that you've got clothes to keep you comfortable or that you can get a glass of clean water anytime you like. In this moment you can choose happiness and gratitude - you can choose any emotion you'd like. This is your moment.

It is a positive contribution to our nervous systems every single time we take a moment for ourselves. It feels great to do it and it adds to our good health. Allow yourself to take a moment whenever you need it and whenever you can. Take a moment before you brush your teeth to smile at yourself in the mirror. Take a moment to think about how much you love your home before you lock the door to leave for work. Take a moment during your walk or drive to think about how fortunate you are to be able to walk or drive. Take a moment to do anything that affirms the health of your mind, body or spirit. This moment will not happen again, so make a habit of adoring your moments where you can. Every single adored moment in your life is one more adored moment in your life. Take a moment

Simply three words

"Just do it" - Nike

It's that time now where we've come to the end of the book. Hopefully you have found many phrases that have touched your heart or helped you to consider things a little differently. Hopefully you have taken more than a few mindful breaths and are fully aware that the power to slow down, relax, make different decisions and to choose happiness all lie completely within your control.

If there are phrases that you feel you still need, I offer that you give the gift of giving those to yourself. Simply three words - they can be whatever you find solace or inspiration in. You're brilliant and know what you need more than anybody else ever could.

Continue your journey to happiness and compassion through mindful moments. Positively change your life simply because you deserve it. Simply three words

If anything in this book helps even one person on this planet to live a happier, healthier, more joyful, fulfilling or peaceful life then every single experience in my life was worth it. Thank you for taking the time to read this and for finding the courage to change. We are all connected and you have given me purpose.

Made in the USA
San Bernardino, CA
29 January 2019